The Studio SOS Book

Solutions and Techniques for the Project Recording Studio

The Studio SOS Book

Solutions and Techniques for the Project Recording Studio

by Paul White, Hugh Robjohns, and Dave Lockwood

Routledge
Taylor & Francis Group

LONDON AND NEW YORK

First published 2013 by Focal Press

Published 2017 by Routledge
2 Park Square, Milton Park, Abingdon, Oxon OX14 4RN

605 Third Avenue, New York, NY 10017

First issued in hardback 2017

Routledge is an imprint of the Taylor & Francis Group, an informa business

Library of Congress Cataloging in Publication Data
White, Paul, 1949–
    The studio SOS book: solutions and techniques for the project recording studio / by Paul White, Hugh Robjohns, and Dave Lockwood.
       pages cm
    ISBN 978-0-415-82386-9 (paper back) — ISBN 978-0-203-54959-9 1. Sound studios.
    2. Sound—Recording and reproducing. I. Robjohns, Hugh. II. Lockwood, Dave, 1956– III. Title.
    TK7881.45.W44 2013
    781.49—dc23   2012050953

ISBN 13: 978-1-138-46889-4 (hbk)
ISBN 13: 978-0-415-82386-9 (pbk)

Cover photo by Chris Capstick

Typeset in Helvetica Neue
by Book Now Ltd, London

Bound to Create

You are a creator.

Whatever your form of expression — photography, filmmaking, animation, games, audio, media communication, web design, or theatre — you simply want to create without limitation. Bound by nothing except your own creativity and determination.

Focal Press can help.

For over 75 years Focal has published books that support your creative goals. Our founder, Andor Kraszna-Krausz, established Focal in 1938 so you could have access to leading-edge expert knowledge, techniques, and tools that allow you to create without constraint. We strive to create exceptional, engaging, and practical content that helps you master your passion.

Focal Press and you.

Bound to create.

> We'd love to hear how we've helped
> you create. Share your experience:
> **www.focalpress.com/boundtocreate**

Focal Press
Taylor & Francis Group

Table of Contents

Preface   viii

1   Monitoring   1

2   Sound Absorbers   20

3   Practical Monitoring Solutions   39

4   The Recording Space   61

5   Soundproofing   77

6   Cables and Connections   93

7   Vocal Recording   116

8   Acoustic Guitars   133

9   Electric Guitar and Bass   147

10   Drums   171

11   Mixing   193

12   Mastering   242

Glossary   261
Index   285

Preface

A decade or so ago, *Sound On Sound* magazine introduced
a new regular column to its pages in which I, Paul White,
Editor-In-Chief and our Technical Editor Hugh Robjohns,
visited SOS readers' studios in person to help solve their
specific recording, monitoring and acoustics problems. *Studio
SOS*, as we titled the column, proved to be a hugely popular
addition to the magazine, and revealed to us many of the
most common studio problems and misconceptions that our
readers were encountering. Taking the concept of hands-
on help a bit further, we subsequently added the equally
successful *Mix Rescue* regular column to the line-up, in which
we take the multitrack files of readers' songs to remix them
ourselves, allowing us to describe the artistic and technical
decision-making processes involved. Between them, those two
columns encompass more or less the entire home and project
studio experience, from setting up the gear to mastering your
album, and over the years that we've been running them, we've
noticed that many of the issues encountered were common to
the majority of the studios we visited. And that's what led us to
this book, gathering all the solutions in one place to serve as the
definitive home and project studio manual.

At one time, high-quality recording equipment was so
expensive that few musicians could afford their own studios,
but now most of us have access to both hardware and
software recording systems that cost less than the price of a
decent guitar. Even the most modest of these is capable of
recording to a very high standard, but all too often the results
are compromised due to inappropriate microphone techniques,
inattention to the recording environment, inaccurate monitoring
or simply a lack of experience. Whilst there are countless
textbooks that purport to tell you everything there is to know
about audio and recording, these sometimes make it difficult
to pinpoint what is really important. The aim of this book is
to allow you to 'cut to the chase' and concentrate on those
areas of recording technique that really make the biggest

difference. So, if you have already started out in recording but aren't entirely satisfied with the results, this is the book for you. It's almost like having your own personal *Studio SOS* and *Mix Rescue* rolled into one!

Often the problems start with the placement of equipment within the room and with the acoustic treatment (or lack of it), so we have some practical, low-cost advice to offer in that area. The next most problematic area seems to be associated with the choice and use of microphones and the acoustic space within which they are used. Once again, these problems can usually be solved very simply and at minimal cost using only the most basic of DIY skills. Once the material has been recorded, even if it has been recorded well, musicians often experience difficulty processing and mixing the parts together to produce the ideal, finished version they hear in their head. This can be due to a number of reasons, all of which we'll explore here: it may simply be due to an inappropriate use of effects or processors, or even a poor arrangement, but there are also several simple mixing techniques that can be employed to make the process easier. Working for *Sound On Sound* magazine we're in the privileged position of being able to interview some of the best music producers, recording engineers and acousticians in the business, and a great deal of what we've learned from them has also found its way into these chapters.

Once you've read *The Studio SOS Book* and put some of its techniques into practice, you'll be amazed by how simple and logical the whole business of recording and mixing is. Your mixes will sound clearer, with front-to-back perspective as well as stereo width, and you may well find yourself far less reliant on radical processing to shape your sound. When you consider that this book costs around the same as a couple of decent-quality XLR cables, it just might be the best recording investment you ever make!

Paul White

Images

Paul White, Hugh Robjohns, Dave Lockwood, Sam Inglis, Chris Korff, Matt Houghton, George Hart

Chapter One
Monitoring

The first and perhaps most important thing to appreciate is that unless you have a monitoring system that is reasonably accurate, you'll never be able to create a mix that translates reliably to the outside world, because you have no means of knowing what your mix *really* sounds like. That may sound obvious, but it is surprising how often we encounter monitoring systems that are quite simply not up to the job. It is worth bearing in mind that even the most expensive professional monitoring systems generate orders of magnitude more distortion and have a far less uniform frequency response than any other element in the entire recording system. Loudspeakers are far and away the weakest link in the chain, and anything you can do to help them work as well as intended has to be worth doing!

Secondly, there's a lot more to good monitoring than simply buying a pair of suitably specified speakers. Both the placement of the speakers and the acoustic properties of the room play a huge part in how accurate, reliable and trustworthy the monitoring system really is. Again, these factors are often significantly underestimated.

When you look at the technical specification of a loudspeaker, what you're seeing is a measurement of the sound projected from the front of the cabinet in an environment where there are no significant reflections from the wall, floor, or ceiling. In other words, you are seeing how the speaker performs in an anechoic test chamber, with around two metres or more of acoustic absorption on every surface, or outdoors, suspended above the ground on a test tower. As soon as you put that speaker into a normal room you will hear the combination of the direct sound from the front of the speaker plus all the reflections of that sound as it bounces off the walls, ceiling

∧

Your speakers and acoustic treatment are amongst the most important elements of your studio – without accurate monitoring you have no means of knowing what your recordings *really* sound like.

and from objects within the room. In addition, you'll also hear any sounds being generated by the speaker's cabinet panels vibrating, (and maybe also the desk on which they are standing) and sometimes air movement noise from any bass ports fitted to the front or back of the cabinet. All these extraneous and inherently very 'coloured' sounds will also be bouncing around in the room and influencing your overall perception of the speaker's sound balance.

What is 'Coloured' sound?

If we look at the sound that we want or expect to hear as the audio equivalent of white light, then anything that changes the balance of frequencies within that sound is the equivalent of a change in the colour of the light – like viewing the world through yellow-tinted glasses. In a similar way, if you cup your hands around your ears you alter the balance of frequencies reaching your eardrums, and the sound becomes coloured. Sound reflections in the room can also compromise the accuracy of the wanted sound from the speakers in several ways, creating inaccuracies both in the frequency spectrum and in the time that different groups of frequencies arrive at the ear – their 'arrival time'. As sound takes a finite time to travel through the air, the reflected sound, which has further to travel, arrives fractionally later than the direct sound, and is also coloured or tonally altered by the reflective properties of the surfaces that it has bounced off. Few materials reflect all frequencies equally, so what bounces back into the room is tonally altered, or coloured by the mechanical characteristics of the wall surfaces.

To compound this effect, most manufacturers only publish the 'on-axis' frequency response, which shows the sound projected directly forward from the speaker. However, the sound doesn't just project forwards in a nice neat 'beam', but actually projects in all directions at once, in varying degrees, and some sound is also generated to a lesser extent by all the surfaces of the speaker cabinet. So every loudspeaker actually has a 'polar pattern' – just like a microphone – and the 'off-axis' sound gets progressively less accurate as the listening position gets further away from the frontal axis. In

▲

Sound reflections compromise the accuracy of the sound from your speakers
in several ways, creating inaccuracies both in the frequency spectrum and in
the 'arrival time' of different groups of frequencies.

particular, the high frequencies tend to fall off and bumps and
dips start to appear in the overall frequency response. In other
words, if you were to sit off to one side of the speaker, what
you'd hear would be significantly less accurate than if you sat
directly in front of it, which is why studio monitors are normally
directed towards the engineer's seat.

In an acoustically reflective room, all these less accurate,
off-axis sounds bounce off the walls and other large surfaces,
and become even more coloured as they are modified by
the characteristics of the reflecting surfaces. Even in the
ideal listening position, what you actually hear is always
a combination of the 'accurate' sound directly from the
speakers mixed with a significant amount of reflected sound
whose frequency response may be very far from flat, with the

reflected sound also arriving at your ears a few milliseconds later than the direct sound.

When those reflections occur within a short time frame – as they do in a typical small-room home studio – our ears are unable to perceive the direct and reflected sounds as separate sources. Thus, our perception of the total sound is determined by the combination of direct on-axis, reflected on-axis, and reflected off-axis elements. Since every listening room will be different in size, shape, construction and contents, it is impossible for a loudspeaker designer to build a monitor speaker that will sound 'right' everywhere: certain assumptions have to be made when the speaker is designed. That's why different speakers can sound so radically different in the same room, even if their on-axis responses appear very similar. However, by using acoustic absorbers to reduce the amount of off-axis sound reflected back to the listening position, we can approach the 'ideal room' acoustics that the speaker designer had in mind, and thus minimize the 'room colouration' – and we'll look at that in more detail in a moment.

Another side-effect of those off-axis reflections is that they come from many different directions, and that can severely confuse or blur the sense of stereo imaging you get when sounds are panned or recorded in true stereo. Fortunately, this problem can be resolved fairly easily with a few strategically-placed acoustic absorbers, as will become evident later.

Resonant Modes

We also need to consider the effect of room 'modes', also known as 'standing waves' or 'Eigentones'. This might sound technical, but what it boils down to is that, in a room with reflective surfaces, the room resonates at frequencies that are dependent on how far apart the surfaces are. There are also further resonances at multiples of those frequencies: if the lowest resonance is at 50Hz, then you'd also expect resonances at 100Hz, 150Hz, 200Hz, and so on. The practical effect of this is to make some bass notes stand out and become dominant or boomy, while other bass notes may sound very weak or even seem to disappear altogether!

There will be independent sets of resonant modes corresponding to the width, the height and the length of the room (these are all called axial modes), and further modes based on sound reflecting between the walls in more complex ways (called tangential and oblique modes). These last two modes, where the sound bounces off two or more surfaces, tend to produce less intense resonances and can actually work in our favour, as they help fill in the more extreme peaks and troughs associated with the primary modes. Rooms with non-parallel walls can spread the distribution of mode frequencies out a little, but it's not the magic fix for room modes that we might hope for.

Unless you build a room where all of the walls act as perfect absorbers down to very low frequencies (if you do, you will have built yourself an anechoic chamber!) there will inevitably

Some Recommended Room Ratios
for Optimal Mode Distribution

	(H, W, L):
Richard H. Bolt, 1946	1.00 : 1.50 : 2.50
Richard H. Bolt, 1946	1.00 : 1.26 : 1.59
The Golden Ratio, 1968	1.00 : 1.62 : 2.62
Dolby Labs	1.00 : 1.49 : 2.31
IEC 60268-13, 1998	1.00 : 1.96 : 2.59

Undesirable Ratios	
Worst ratio (RPG Inc.)	1.00 : 1.07 : 1.87
Cube - Not recommended	1.00 : 1.00 : 1.00
Exact multiples	2.00 : 1.00 : 1.00

Acceptable room size ratios determined by Bolt

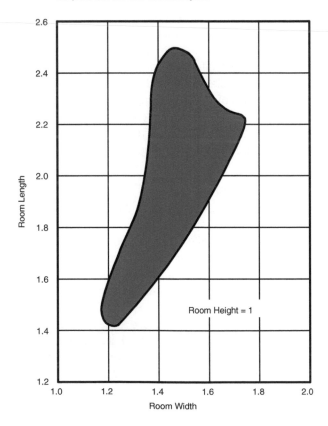

Room Height = 1

Room Length (y-axis)

Room Width (x-axis)

be resonances due to room modes in your studio. This is just a fact of life, so rather than fight the room modes we have to try to design our studio to work with them. A very important fact to keep in mind is that in a good-sounding room, the resonant modes will be spread around as evenly as possible to effectively average out the resonant response of the room. In a poor-sounding room, the modes will be unevenly spaced, with several bunched together around specific frequencies. And where there's bunching, you'll hear a corresponding increase in perceived level.

To achieve an even modal response, the room needs to have dimensions that are not exact multiples of each other. Not surprisingly there has been a lot of research into this and, while there is no absolute perfect set of dimensions, several optimal ratios of room width, height and length have been calculated, most notably by Richard H Bolt in 1946.

Room Shape and Size

It is also a factor that the larger the room, the more room modes you have. This is because low-frequency audio has very long wavelengths – 60Hz has a fundamental of just over 5.7 metres (roughly 19 feet) – resulting in some multiples simply not fitting within the dimensions of smaller rooms.

In a larger room, with more modes, the chances are that they will average out to something close to a nominally flat response. In contrast, a small, cube-shaped room with very solid walls is pretty much the worst-case scenario: because the room is small, the room modes are fewer in number and so more widely spaced in frequency, and because all three sets of dimensions are the same, the resonant modes for all three sets of surfaces will occur at the same frequencies. This creates huge resonant peaks at some frequencies where the reflected energy combines constructively, because it is in phase, but there will also be very deep notches or dead spots at other frequencies where the reflected energy cancels out because it combines out-of-phase. The practical outcome of trying to mix in rooms with dominant modes is that the level of your bass line will appear to vary dramatically depending on which note is played, even though they may really all be at the same level. Music in one key might sound fine, while other keys could sound terrible! Experimenting with the speaker placement can often improve the situation (by not exciting the major modes so easily), as can installing purpose-designed 'bass traps'. The lightweight walls of modern houses are a big help too, as they allow the bass to effectively 'leak out' of the room – but we always know we're in for a tough day if we come across a small, cube-shaped room.

Another feature of small, cube-shaped rooms that we encounter often during our *Studio SOS* adventures, and one that I haven't seen mentioned in many other recording books, is that if you place your head in the exact centre of the room the overall perceived level of bass drops off alarmingly. This seems to occur in a roughly spherical 'void' of 0.5–1 metre across, prompting us to nickname it 'The Spherical Bermuda Triangle of Death!' Thankfully, only the music 'dies' there, though – no humans have been harmed as far as we know!

Unfortunately, many small, bedroom studios are very close to cube-shaped, and by the time the equipment desk has

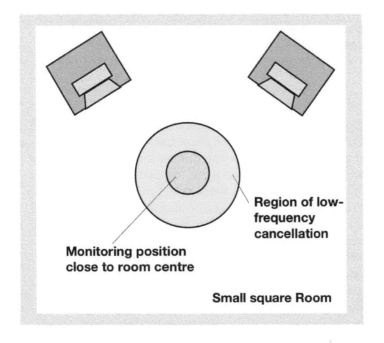

Region of low-frequency cancellation

Monitoring position close to room centre

Small square Room

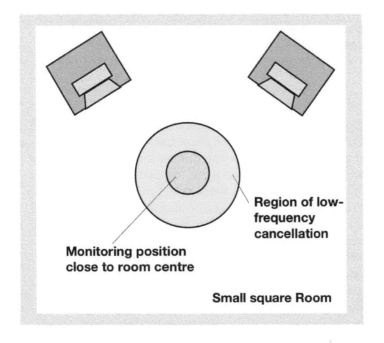

In a very small, square room, there is sometimes nothing you can do to avoid sitting in the middle, in the area where low-frequency cancellation occurs.

been set up, the engineer's chair inevitably ends up being very close to the centre of the room, and this is compounded by the engineer's head being almost midway between the floor and ceiling. This is the absolute worst monitoring situation and should be avoided if at all possible, as you'll get a very misleading impression of how much bass end you have in your mixes. We have some tips for working in less-than-ideal rooms, but these do inevitably involve a degree of compromise.

Another unfortunate room shape is one in which one set of dimensions is an exact integer multiple of one of the others: a room that is twice as long as it is wide is a prime example. While not quite as disastrous as a cube, you'll still find resonances stacking up at the same frequencies to create hot spots and dead spots in the low frequency spectrum. In this situation there is probably a better chance of being able to arrange the room to get away from any dead spots created in the exact centre, but having unrelated width, height and length dimensions is always the best option (See box and diagram for recommended room dimension ratios).

Note that in a room with a lightweight suspended ceiling of the types used in offices, the modal behaviour of the room will depend on the original, structural roof ceiling height, as the false ceiling will be virtually non-existent to low-frequency sound.

Checking for Bass Problems

While a professional studio designer will use sophisticated measuring equipment and software to determine how the room response behaves, such software can be misleading in inexperienced hands. We have found that playing a chromatic sequence of staccato sine-wave tones, all at equal level running for around three octaves from around 250 Hz (C4) down to 31Hz (B0), can be very revealing in a more intuitive way. This can usually be done quite easily within most DAW software, using either a software synth, or an oscillator plug-in (see p. 18 – Making A Sine Wave Test Sequence). All you have to do is run the sequence while listening from your usual mix position: any serious bumps and dips in the room's frequency response will be clearly audible as individual notes in the series being too loud or too quiet in comparison with their neighbours. The speaker positioning can then be adjusted, as described later in this chapter, to try to minimise these irregularities. If the room is behaving correctly, the overall level will appear to fall away as the frequency gets lower (because of the loudspeaker's own low frequency roll-off) but no individual notes should stand out as being particularly louder or quieter than the adjacent notes.

Across or Along?

In large commercial studios, you'll often find that the room is wider than it is from front to back. However, if space is limited – and in home studios it invariably is – you'll almost certainly get better results by arranging the monitors to point down the long dimension of the room rather than across it. In virtually every case we've tried, having the speakers firing down the length of a small or medium-sized room has produced a more even bass response than firing across it.

/ CASE STUDY – PAUL WHITE

My own studio is roughly 16 feet long, 11 feet wide and 7.5 feet high. I've tried orientating the room and monitors both ways and the difference is remarkable. With the speakers firing across the room, the perceived level of bass changed quite drastically as I moved my listening position even slightly, and there were very noticeable cold spots where some notes were much quieter than others. This effect is partly due to the fact that when working across the room, the listening position ends up being roughly mid-way between the front and rear walls where the most significant bass cancellation effects occur. The practical outcome was that I was unable to make any meaningful judgment about how the bass end of my mixes really sounded.

"By contrast, working with the monitors firing down the length of the room, as I do now, shows up no such problems, and the perceived level of bass stays pretty constant regardless of where you are in the room, unless you stand very close to a wall or corner. The 'bass boost' in these positions is to be expected because low frequencies always build up in the 'pressure zone' close to walls or corners due to the way the direct and reflected energy combines."

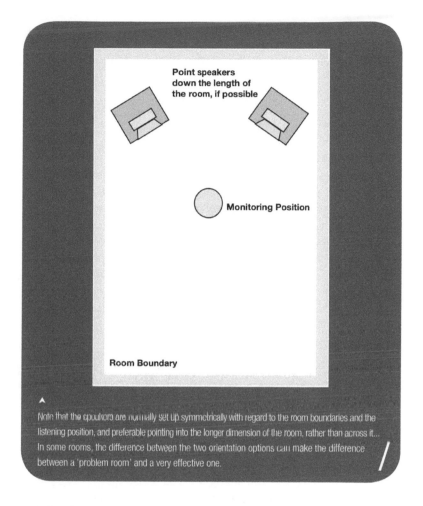

Point speakers
down the length of
the room, if possible

Monitoring Position

Room Boundary

Note that the speakers are normally set up symmetrically with regard to the room boundaries and the listening position, and preferable pointing into the longer dimension of the room, rather than across it... In some rooms, the difference between the two orientation options can make the difference between a 'problem room' and a very effective one.

Monitor Placement

To create a correctly symmetrical stereo image, your monitor speakers should, ideally, be set up symmetrically within the room. Although this isn't always possible in a domestic environment, you should aim initially to get as close to true symmetry as is possible. In other words, your speakers should be the same distance from both of the side walls while, the distance between each speaker and the wall behind them should also be the same. You should also try to arrange any furniture in the room as symmetrically as possible so that the reflections are similarly balanced. Theory also dictates that if you can place the speakers at a point 38% of the room's length from one end, you'll get the smoothest bass response, though in most real-life situations, space constrains means that the speakers usually have to be rather closer to one end of the room.

Asymmetry Can Help

In very small rooms, or rooms where other factors have meant the speakers have to fire across the room, we have found through our *Studio SOS* visits that there can often be a benefit in breaking the 'perfect symmetry' rule slightly. Shifting the line of symmetry a few inches left or right of centre – which means moving the two speakers, the listening position and the whole setup a little way left or right of where they should be in a perfectly symmetrical set up – can often make a worthwhile improvement in the linearity of the low frequency

TIP: In a small room it is sometimes impossible to get the speakers as far away from corners as you would like, but it helps considerably if you arrange the speakers so that the distance between them and the side wall is different from the distance between the speakers and the wall behind them. If possible, also avoid setting the speaker height so that the woofer (bass driver) is exactly midway between the floor and ceiling.

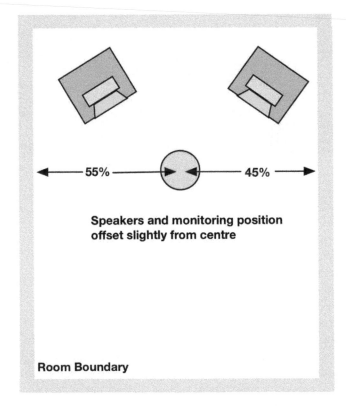

55% — **45%**

Speakers and monitoring position offset slightly from centre

Room Boundary

◄

Although speakers are normally set up symmetrically with regard to the room boundaries, in a small room, moving both the speakers *and* the listening position slightly off centre can sometimes make a worthwhile improvement in the evenness of the low frequency response. Here, the monitors have been offset slightly to the right.

response. This is because the speaker-to-wall dimensions are now different for each side, which affects the efficiency with which they are able to energise standing waves at different frequencies. Changing the distance of the speakers from the wall behind them a small amount may also help.

The Magic Triangle

The accepted optimum monitoring geometry is to set up compact, or 'nearfield' speakers (which most home and project studios use) so that they form two points of an equilateral triangle with the listener at (or slightly in front of) the third point. In most cases you should angle the speakers inwards so the high-frequency drivers ('tweeters') point directly at, or slightly behind your head when you're seated in your normal listening position, and if you have to tilt the speakers slightly to achieve this in the vertical plane, that's fine too. However, check what the manufacturer's handbook has to say about positioning and the angle (often referred to as the toe-in angle) as a few speakers are designed to sound most balanced when aimed well behind the listening position or even straight down the room. As a rule, the speakers will end up fairly close to the front wall in a small room but try to keep them at least a 30cm (12 inches) away from it if at all possible. This is particularly important for speakers that have ports or passive radiators on the rear panel as being very close to a wall can compromise their operation.

Stands

Where possible, put the speakers on rigid floor stands. Stands are often made from tubular metal and filling this type with dry 'play-pit' sand, lead shot or 'kitty litter' will make them more stable and also damp out any resonances in the metalwork. A pea-sized blob of the type of putty used for fixing posters to walls (such as Blu Tak) under each corner of the loudspeaker reduces the risk it being knocked off the stand – just allow the speakers to compress the putty as far as their weight allows – and this also provides some useful vibration decoupling between the speaker and the stand.

◄

Where possible, speakers should be mounted on rigid floor stands, elevating the speakers so the tweeters are around ear height for a seated listener.

Where the speakers must be placed directly onto a desk, on a shelf, or on the meter bridge of a mixing console, you will almost certainly get better results if you use commercial high density foam supports to isolate them from the desk – characterised by a firmer, tighter bass sound. These sometimes also come with additional foam wedges that can be used to adjust the vertical angle of the speakers. The foam helps prevent vibration from the speakers getting into the structure they are sitting on and causing audible resonances. As a rule we wouldn't recommend this type of foam decoupling for use with rigid speaker stands, but for table-top or shelf mounting, they have produced consistently good results.

Some speakers, however, are so small and light that they don't compress the foam sufficiently for it to absorb low frequency energy very efficiently. The foam needs to be

Speakers placed directly onto a desk or shelf can sometimes benefit from the use of high-density foam pads to provide a degree of decoupling. The Genelec monitor on the right has its own built-in Isopod feet, designed to do the same job.

compressed sufficiently to allow it to behave as a 'damped spring'. In such cases we have found that placing a heavy ceramic tile between the foam and speaker often does the trick. There are also commercial speaker platforms with heavy metal plates fixed on top of the foam to work in the same way and these also work very well, producing a worthwhile increase in low-end clarity.

Adding a high-mass plate, as used in these Recoil Stabilizers from Primeacoustic, ensures that the foam underneath will act as a 'damped spring'.

Room Reflections

Having dealt with monitor positioning, it is also very important to examine more closely the influence of room surfaces on the sound and to take appropriate measures to minimise unwanted reflections arriving back at your listening position. When sound hits a hard, flat surface, it bounces off it much like light from a mirror, and just as a light bulb viewed in a mirror emits its own 'phantom' light source, a hard surface that reflects sound from a speaker produces a similar effect, just as though there was a phantom loudspeaker behind the wall. The best way to understand this is to get somebody to hold a mirror flat against the side walls of your studio room while you sit in your mixing chair. When the mirror is in a position that allows you to see the monitor speakers, you know that sound reflections from that point will also bounce back to your listening position. The sonic effect is very much the same as if the speaker you saw in the mirror was real and emitting the same sound as the main speakers, but very slightly later and with a different tonal balance, as we've already explained – and these reflections will compromise the accuracy of your monitoring and confuse the stereo imaging. While we can simply paint a mirror black to stop it reflecting light, devising an audio equivalent of black paint isn't quite so straightforward.

Note that in a typical domestic room, the dimensions are such that any reflections from walls, the floor and the ceiling will arrive back at the listener so quickly that they blend in with the original sound rather than being perceived as separate echoes. The effect of this is that the overall sound appears to be coloured – there's no sense of hearing accurate direct sound alongside delayed coloured sound. Effects due to multiple reflection paths are audible mainly at mid and high frequencies and are fairly easy to deal with, whilst modal problems related to room geometry result in mainly low-frequency problems.

Monitoring accuracy also suffers if speakers are set up with a large area of reflecting surface, such as a table top, directly in front and below them. This is because both direct and reflected sound reaches the listener causing phase cancelation problems, resulting in peaks and dips in the mid and upper frequency range. If the speakers are set up on floor stands or a shelf, aim to keep this as high above the desk surface as possible and try

TIP: Don't allow anything to come between you and the loudspeaker, such as the edge of a computer monitor screen. Sound reflections from such surfaces compromises accuracy and degrade the stereo imaging quite dramatically, especially if the tweeter becomes partially obscured. Ideally, the computer monitor screen, and anything else placed between the speakers, should be positioned well behind a line drawn between the speakers' front baffles.

TIP: Don't lay speakers on their sides unless the manufacturer's instructions confirm that they are specifically designed to be used that way. Normally the tweeter should be directly above the woofer so that the high and low frequencies remain correctly time-aligned as you move to the left or right of the normal mixing position at the apex of the equilateral triangle we mentioned earlier. Using such speakers horizontally can destroy that time-alignment and will not only significantly narrow the monitoring 'sweet spot' within which the mix sounds accurate, but it will also result in comb-filtering colouration if you move from side to side as you work.

If the tweeter is mounted above and to one side of the woofer, it may be an indication that the speaker has been designed to work in either orientation, and some dual-woofer speakers are also designed to work horizontally, with the tweeter located between the woofers. However, you should always check the manufacturer's instructions for clarification as other design factors may influence the mounting options.

▲

When looking for reflection paths, don't forget the surface you are sitting at! Angling the speakers up and away from the table surface, or adding something absorbent can make a significant difference to the accuracy of what you are hearing.

to angle the speaker to minimise the amount of sound projected down to the table surface. Where the speakers must sit on the desk itself, move them as close towards you as is practical and angle them upwards so the sound is directed towards your head and not towards the table top. Minimising the size of the desk can help too, especially if you can work with one that is narrower than the space between your speakers so that you can place speaker stands at the sides of the table instead of directly behind it. In extreme cases, we've had success placing one- or two-inch-thick acoustic foam panels on the desktop in front of the speakers to help kill reflections. If you're installing acoustic foam in other parts of the studio, you may have some off-cuts that can be used for this purpose. Another viable alternative is to use a perforated metal computer desk so that most sound will pass through it rather than bounce off it.

So much for the problems, it's time now for some practical solutions.

/ MAKING A SINE WAVE TEST SEQUENCE

A very simple and intuitive way to identify and assess uneven bass problems caused by room modes is to play a chromatic sine wave test sequence and listen for which notes stand out as too loud or too soft – and here's how to set it up. It may be worth burning the results to an audio CD to make future tests easier to carry out.

Using an oscillator plug-in, a software synth, or a software sampler (some play sine waves by default until you load a sample) in your DAW, program a MIDI sequence to play a pure sine tone over the bottom couple of octaves starting at C4 (250Hz) and running in semitone steps down to about 55Hz (low A0) or even 41Hz (low E0). The lowest fundamental of a five-string bass is 31Hz (low B0) and very high quality speakers in a large room might be able to reproduce that, but few compact two-way speakers will be able to generate much energy at such low frequencies, so in most cases there's little point in taking the sequence much lower than 41Hz.

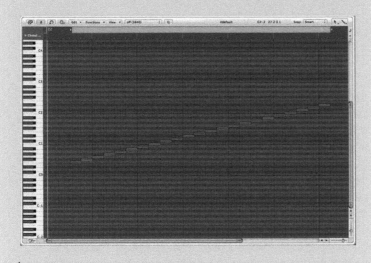

▲
Chromatic sine-wave test sequence created in DAW software.

Each note in the sequence needs to be about one second in duration – long enough for any standing waves to have time to build – but with gaps between notes so that you can hear any modal ringing effects when the note stops. Make sure all the notes have the same MIDI velocity, and then arrange for them to play in a ascending sequence. At the starting higher frequencies the levels of adjacent notes should remain very consistent as they are above the frequency range usually affected by room modes, but as the sequence descends you will become aware of some notes being louder and boomy, and others being very weak or even absent altogether!

As the sequence plays listen to the relative loudness of each note, both from the listening position and as you walk around the room. If they sound more or less even you're in luck but, if not, try moving the speakers by a few inches and try again, and if that doesn't help, consider installing bass traps (see Chapter Two).

Chapter Two
Sound Absorbers

The ideal recording or monitoring room is one where the reflections and reverberation of all frequencies decay fairly quickly and at a similar rate, so that the sound remains tonally balanced. In a room with a lot of hard, reflective surfaces – wood, glass, tiles, polished concrete, etc. – the high frequencies bounce around a lot longer than the mid and low frequencies, and we perceive the space as being 'bright-sounding'. Conversely, a room with a lot of carpet and soft furnishings, where the high frequencies are absorbed and therefore decay more rapidly than the low frequencies, will tend to sound warmer or perhaps even rather dull. The fundamental aim of acoustic treatment, therefore, is to try to control the sound reflections in a room in a way that maintains a good balance across the whole frequency spectrum.

Whilst a professional studio installation will incorporate some very sophisticated (and usually expensive) acoustic treatment and involve a lot of calculations and measurements, the reality is that even the sometimes very basic acoustic treatment we can install during a typical *Studio SOS* visit brings about a vast improvement when compared with the original untreated room.

Absorbing Sound

There are numerous ways of dealing with sound reflections in a room, the majority of them using what are known as porous absorbers – a porous absorber is one that allows air to pass through it, absorbing some of the sound energy from the air as it does so, leaving less energy to be reflected

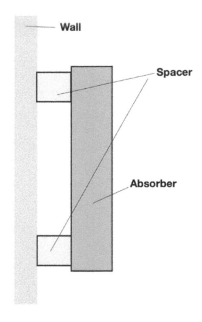

Wall

Spacer

Absorber

The efficiency of an acoustic absorber can always be improved by spacing it off the wall a little.

TIP: At 1kHz, a quarter of a wavelength is approximately 86mm while at 100Hz it is 858mm, assuming that sound travels at around 343m per second (it varies slightly with the ambient temperature and humidity).

back into the room. Popular examples of porous absorbers are acoustic foam, mineral wool, glass fibre and woven blankets. However, the important thing to note about all of these materials is that the thicker they are, the better they are able to absorb low-frequency energy – because lower frequencies have longer wavelengths. Their effectiveness is also improved if they are spaced slightly away from their mounting surface rather than stuck directly to it, although where space is limited, fixing the absorbers directly to the wall will still be of value. Some commercial foam products even come with additional foam spacing blocks for this purpose. The technical explanation for the preference for spacing arises from the fact that there is no actual air movement immediately in front of a reflective surface, just variation in pressure, and, of course, an absorber that works by taking acoustic energy out of the moving air only works well when air is actually trying to move through it! To achieve maximum effectiveness with a porous absorber, you would locate it a quarter wavelength from the wall, as that is the region where the air molecules are moving most.

Although the physics behind all of this can seem rather complicated, the practical outcome is that to be optimally effective absorbers need to be a quarter wavelength thick (or more) for the lowest frequency you are trying to absorb. When you consider that the quarter wavelength of a 50Hz bass note is around 1715mm you can see you'd need a five-foot thickness of absorber to be effective, and although you could get worthwhile results by using a lesser thickness spaced five feet from the reflective wall, you'd lose a lot of floor area which ever approach you take! Of course, some reflected sound hits the absorbers at an oblique angle and therefore travels through a greater thickness of the material before reaching the reflective back wall. The practical outcome of this is that a given thickness of absorber still offers some useful attenuation below the quarter-wavelength frequency to off-axis sounds.

Clearly then, purely porous absorbers are not entirely practical as low-frequency absorbers, although there are ways of deploying them to make them more effective than you might imagine without filling your entire studio with foam or mineral

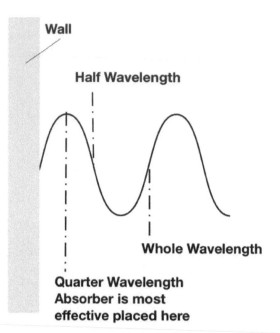

Wall

Half Wavelength

Whole Wavelength

**Quarter Wavelength
Absorber is most
effective placed here**

To be optimally effective, an absorber needs to be at least as thick as a quarter wavelength of the lowest frequency that you want to attenuate.

wool! Shallower acoustic absorbers can be very effective at mid and high frequencies and are very easy to implement, and there are other, more practical, strategies for helping to deal with the low end.

A panel of acoustic foam or mineral wool around 100mm thick and spaced 50mm or so from a wall gives useful absorption down to around 250Hz, although it is most effective above 500Hz. A 50mm foam or mineral wool panel fixed directly to a wall, by contrast, will start to lose effectiveness at somewhat higher frequencies and is only useful down to 750Hz or so, with the greatest effect being above 1.5kHz. While such absorbers do little to help control bass problems, they are vitally important in controlling those mid/high reflections that would otherwise cause colouration and blurring of the stereo image.

A number of companies make acoustic foam with profiled surfaces in the form of wedges, waves or pyramids. These probably make the foam more effective at intercepting sound approaching at an angle but they also reduce the average thickness of the foam and thus affect its efficiency at lower

>

Off-axis sound actually travels through the absorbent material for a greater distance and is thus attenuated slightly more than directly incident sound.

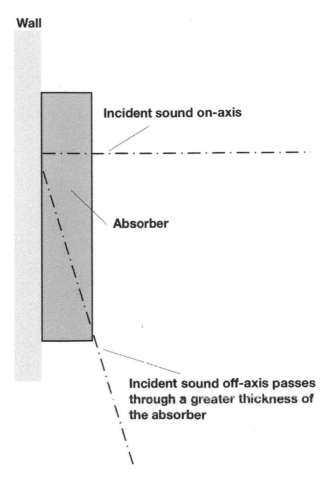

Wall

Incident sound on-axis

Absorber

Incident sound off-axis passes through a greater thickness of the absorber

frequencies. However, if you can space these off the wall, the impact on low-frequency effectiveness will be lessened and the overall efficiency of the absorber improved.

Where you don't want to glue the foam permanently to a wall, you can attach it to a ply or MDF panel and hang it as you would a picture frame. However, if you choose a heavily perforated board (such as the type designed to cover domestic radiators) rather than a solid one, you can space the panel around 50mm from the wall using wooden or foam blocks behind, to improve its low-frequency efficiency. (There would be no advantage in spacing it from the wall if you used a solid backing board as that wouldn't allow sound waves to pass through.)

TIP: While there are no acoustic problems in mixing and matching different thickness acoustic foam tiles from different manufacturers, beware that they are not all made to the same dimensions. Most American-made tiles are 2ft square, while European-sourced designs tend to be 600mm square. The difference is roughly 1cm in each dimension!

◄
Acoustic foam panels often have a profiled surface to increase their efficiency at high frequencies, at the expense of some low frequency performance.

If spacing the foam away from the wall is not essential, as may be the case with thicker foam, you can also hang it 'picture-style' by gluing an old CD to the centre of the top rear edge and then hanging the hole in the CD over a nail or screw placed in the wall. This is a good solution for working in rented accommodation as gluing directly to the wall invariably damages the wall surface when you come to remove the tiles. We've also used the picture rails in those older houses that still have them, by hanging the panels from metal hooks designed to hang on the rail's profile.

A commonly used alternative to acoustic foam is mineral wool; the rigid type sold for cavity wall insulation (often in 600 x 1200mm slabs, 30 or 60mm thick) is both inexpensive and very effective as acoustic treatment. Multiple layers

TIP: If gluing acoustic foam tiles to walls, be careful to use a spray adhesive specifically designed for the job. Many forms of adhesive will dissolve foam, destroying your expensive acoustic tiles in seconds!

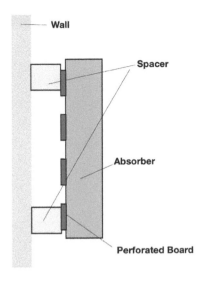

If you use a backing board to space foam panels off the wall, the board must be perforated, otherwise the foam will perform just as if it were mounted directly on to the wall.

can be used to build up the desired thickness, but even a single layer of 30mm material makes a useful mid/high trap if spaced away from the wall. In theory this dense 60kg per square metre grade may not be quite as efficient at absorbing higher frequencies as the slightly less dense 45kg per square metre variety, but we've found it to be adequate in practice and it is much easier to handle than the less compacted forms.

On several of our *Studio SOS* missions we've built traps that combine inexpensive mineral wool, to provide the bulk of the absorption, with a thin acoustic foam panel mounted on the front to provide a more cosmetically acceptable finish, and to prevent mineral fibres from breaking free. The foam also adds a little more thickness to the absorber to extend its effective bandwidth slightly. This type of trap is affordable, looks very professional, and is very simple to make. Our design utilises a simple wooden frame 100 to 150mm deep with the mineral wool slab set back by a few millimetres

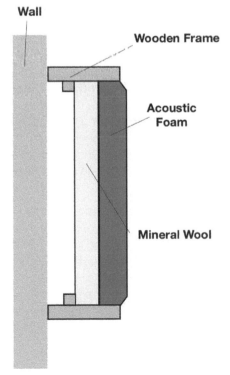

A good-looking, easy-to-make absorber can be created by layering an acoustic foam tile on top of a slab of mineral wool.

from the front edge. The mineral wool is held in place by wooden battens pinned into the frame directly behind it; this arrangement leaves a useful air gap behind the mineral wool. The acoustic foam is then glued directly onto the mineral wool using a light spray of the contact adhesive supplied for fixing foam tiles and, because the mineral wool is set back a little way, the edges of the foam can be tucked inside the frame to give a neat finish.

A more cost-effective DIY option for those on a tighter budget is to use the same mineral wool slab fixed in a wooden frame again, but this time positioned flush with the front edge, and leaving an air gap of 25 to 75mm behind it. The whole frame can then be covered with a thin, porous cloth, held in place using staples around the back of the frame.

Note that mineral wool should be handled carefully to avoid exposure to the loose fibres that can cause irritation to the lungs and skin to some people. Spraying the surface lightly with a very dilute (1:4) mix of PVA glue and water can help to prevent shedding of surface fibres without affecting the acoustic performance of the mineral wool. Commercial absorbers are often covered with an acoustically transparent fabric called 'Cara', but while this looks very good and is available in many colours, it can be quite expensive. The cheapest suitable covering we've found is the black, weed-control, treated paper sheet available from garden centres. It looks surprisingly good and is completely breathable, although it isn't very tough for use in high traffic areas.

Some commercial acoustic panel manufacturers now prefer to enclose the mineral wool in very thin polythene bags prior to use. Most of the sound will still be absorbed by the mineral wool because the thin polythene is able to move slightly when sound waves hit it, recreating the air movement from one side on the other where its energy is absorbed by the mineral wool. Although a small amount of very high-frequency reflection also occurs from the polythene, most of that will be taken care of by the covering fabric.

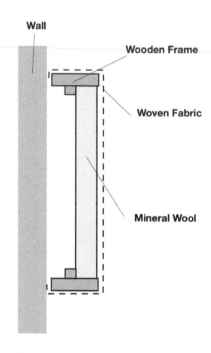

A simple mineral wool absorber can be given a professional look by attaching a breathable fabric covering to the frame.

Carpet, Egg Boxes and Other Myths

From what we've said about the relationship between thickness and the lowest frequency that can be absorbed by a porous absorber, you'll now see why sticking carpet directly to walls in the hope of providing cheap and attractive sound absorption is just not going to work. Carpet is too thin to be effective at frequencies lower than several kHz. When a studio is lined only with carpet on the walls, the carpet absorbs the very high frequencies but leaves the midrange and low end completely untreated. That, in turn, leads to the mid- and low-frequency reflections dominating the acoustic of the room, making it sound boxy and dull. Remember, in our ideal room all frequencies should decay at an approximately equal rate.

In fact, we have visited a number of studios treated with carpet glued directly to the walls and they have all exhibited a dull, boxy sound, just as theory predicts. The only way to rescue such a room without stripping away all the carpet and starting again is to add some high frequency reflective material to try to rebalance the tonality of the room, and if possible introduce some additional mid- and low-frequency absorption to help dry up the lower. Any thin, rigid hard material will reflect high frequencies, so carpeted rooms can be rescued by the application of some MDF, glass or perspex panels, in conjunction with some more effective mid-range mineral wool or thick foam absorbers.

However, if you have a lot of spare carpet, or particularly like the padded-cell look, you could use a thin, porous carpet (not a rubber-backed one!) to provide an acceptable cosmetic front to DIY absorbers based on mineral wool or other suitable material, mounted on the walls. We've seen very attractive and effective studio acoustic treatment where entire walls have been covered in deep mineral wool absorbers, which were hidden from view by being covered with a porous carpet to give a neat, hardwearing acoustic wall!

Carpeted floors, however, are not particularly problematic – and we'll get to egg boxes later!

CASE STUDY – CD DISTRIBUTION!

Picture frames with glass fronts, or mirrors, combine useful acoustic reflection with decoration, and this is something we used when fixing up one carpet-treated studio: in this instance, the walls of both the control room and live performance space were completely covered in thin cord 'contractors' carpet. However, a few picture frames were insufficient to produce a large enough reflective surface to redress the balance so additional hard surfaces needed to be added. We've used the thin, rigid plastic-foam panels designed to be used beneath laminate flooring to good effect, and also thin MDF sheets. We've even had success sticking old CDs inside over-absorbent vocal booths to bring back some high-frequency liveliness. Once you've correctly analysed what the problem is, it is usually fairly easy to improvise a solution, as any hard material will reflect high frequencies reasonably well. The trick is to spread out the reflecting panels rather put them all in the same place, as you may need to cover up to 50% of the wall surface to put some life back into the room.

A small vocal booth with too much high-frequency absorption can be brought back to life with some improvised distributed reflective surfaces – unwanted CDs are ideal!. We taped them to the wall as a temporary measure to find out how many it would take to do the job and where best to site them!

Decay Time

As we have established, most good-sounding rooms have a decay time that is nominally similar at all frequencies, although it is acceptable for the low-frequency decay time to be slightly longer as very low frequencies are difficult to absorb. Technically, the term 'reverberation' (or 'reverb') isn't really applicable to very small rooms as the sound reflections never manage to build up into true reverberation, but the various resonances within the room do mean that different frequencies take different times to decay.

The average decay time for a studio space across all the frequencies will vary enormously depending on the size of room and its intended application. A concert hall will have a fairly long reverb time – typically between one and two seconds, designed to flatter the sound of an orchestra – whilst a contemporary music studio will have a much shorter decay time that sounds 'live' without having any dominant resonances (usually less than one second). A studio control room, or radio/television voice-over booth, by contrast, will have a very short decay time, perhaps around 0.6 seconds or less. The decay time of a reverberant space is often specified as its RT_{60}, which is the time it takes for a sound to decay in level by 60dB.

Less Effective Solutions

Other near-useless 'absorbers' that we have seen set up in studios we've visited include expanded polystyrene panels, which actually reflect high-frequency sound, and canvas-covered mineral-wool absorbers where the canvas has been painted for decorative purposes making it more or less airtight! You'll remember that an absorber can only work if air can move through it, and whilst painted canvas will still pass some sound, its heavy weight means that the performance of the absorber will be much less efficient and quite unpredictable, with the painted surface likely to reflect more high frequency energy than is desirable.

Open-plan-office dividing screens can also seem an attractive, ready-made solution for general acoustic treatment, or for screening off instruments to gain better separation when recording. However, most of these are designed to offer only modest absorption, primarily at voice frequencies – mainly above about 2kHz. Their effectiveness can, however, be increased by draping heavy blankets or duvets over them, or you could simply glue acoustic foam panels directly to one side.

Then there's also the old egg box myth! Egg boxes don't help at all with soundproofing (which is a very different subject from acoustic treatment and will be discussed later), and their rigid surface doesn't do much in the way of absorbing either. Their shape makes them marginally useful as diffusors (discussed in the next section) but the indentations are too small to make them effective even for this purpose other than at fairly high frequencies. There are far better solutions for diffusing sound, and let's face it, egg boxes don't look very professional!

Flutter Echo

Clap your hands when standing between two closely-spaced parallel walls with untreated hard surfaces and you'll hear a noticeably resonant ringing sound following the handclap. That is flutter echo in action! In addition to reflecting sound from the speakers back to the listening position, sound will also bounce back and forth between the walls. Fortunately, it is relatively easy to deal with flutter echoes by using absorbers, as already discussed, or by creating a deeply irregular surface that scatters the sound equally in all directions, rather than directly back where it came from.

Scattering

When deliberately introducing high-frequency reflective areas it is preferable not to use large flat surfaces parallel to the walls. This is because their influence on the sound, and particularly on the stereo imaging, can be disruptive, especially if they are close to the monitoring position. Although multiple small flat panels are OK for a quick fix, it is usually better to use areas of curved (convex), angled or irregular material if possible as this helps to diffuse or scatter the reflected sound rather than bouncing it straight back like a mirror. Pursuing our mirror analogy, you can think of diffusion as replacing a normal flat mirror with a surface made up from irregularly fixed mirror tiles at different angles, forcing it to reflect in a random way. We have successfully used old CDs as small reflectors, but you can use thin wood strips, or small plastic or metal items, as the aesthetic requirements dictate. Split logs can be good too, as their curved surface scatters sound very effectively. There are also, of course, a range of elaborate commercial diffusers that work well and look great, but they tend to be quite expensive. It is possible to build your own diffuser by gluing wooden blocks of different lengths to a back board, but a shelf full of gear, books and CDs often works almost as well – provided they are randomly distributed – and may be a better use of space in the smaller studio.

Commercial studio designs may use angled wall panels as well as non-parallel walls and shaped ceilings to divert

> The distribution of shapes and spaces in a commercial diffuser is carefully calculated for optimum results, using complex mathematical formulae, but that doesn't mean you can't achieve worthwhile results with something a bit more DIY.

reflections away from the listening position. As a rule though, this only works well in large rooms and involves a lot of design expertise and expensive building work. Most of us have to make do with domestic rooms or converted garages with parallel walls.

Bass Traps

Low frequencies are the hardest to deal with, because of the relatively high acoustic energy involved and their long wavelengths. Simple absorbers generally need to be very deep to deal with low frequencies effectively, as we have already explained, which isn't usually very practical. However, it is possible to use corner absorbers to help smooth out the lows without taking up too much floor area, and we'll come back to that technique shortly.

If you look at traditional acoustic design textbooks you'll often find so-called 'tuned' traps recommended for controlling room modes, and these are essentially tuned resonant chambers designed to absorb only around a specific narrow range of low frequencies. If you have only one or two prominent room mode hot spots then this can be a very effective solution, but these traps are notoriously difficult to get right as any variation in dimensions or material specifications can result in them absorbing at a frequency outside the range you wish to deal with. We have yet to use tuned traps on a *Studio SOS* job as we generally have only one day to address the problems we find.

Broadband Bass Traps

The simplest bass traps to build for the home studio owner are broadband traps using simple absorbing material, or a combination of absorbing material and 'limp-mass' absorbers (see below). These can be effective, and because they are broadband, no calculations are involved. The principal low-frequency room modes are 'anchored' in the corners of the room making the space across corners the most effective place to put bass traps.

You may have seen commercial foam wedges that fit into corners and these can be useful when the problem isn't too severe. They can be placed in any corner but, as with all acoustic treatment, it is best to keep the layout reasonably symmetrical from left to right, if possible, as they will also absorb mid- and high-frequency sound. If the vertical corners of the room are unsuitable for whatever reason, you can also utilise the corners between the walls and ceiling or even the wall/floor corner behind your mixing desk. Ideally the traps should run the full length from floor to ceiling or, if fitted to the wall/ceiling junction, across the full width and/or length of the room. The thicker and larger the foam used to form the corner trap, the lower the frequency to which it will be effective, but the length of the trap is arguably equally important as this design relies on a long length of trap presenting a greater 'depth' of foam to waves approaching from an angle rather than simply square-on.

Commercial foam bass traps can be expensive but a DIY alternative is to build a frame that holds a standard 600 x 1200mm high-density mineral wool slab diagonally across a corner, then fix several of these across the corners to fill the desired space. The triangular void behind the traps can then be filled with lower density mineral wool or standard loft insulation – but acoustic foam offcuts or even old blankets will help (but ensure you use fire-retardant materials for the sake of safety). If you have some old furniture foam you'd like to use, check that it is porous by trying to blow through it. If you can blow through it fairly easily, it is porous. If not, it may be the type with sealed cells rendering it useless for acoustic treatment.

Rooms with unused alcoves, fireplaces or ceiling voids may provide you with space to add deeper absorptive traps made from compressed mineral wool. In the bedroom studio you can sometimes get a useful amount of 'free' bass trapping by storing unused bedding on top of wardrobes and also by leaving the doors of well stocked wardrobes open. Mattresses and sofas also provide a degree of useful trapping, though you need to let your ears be the judge if it is enough. We've even created instant bass trapping by stacking rolls of loft insulation in room corners, still in the

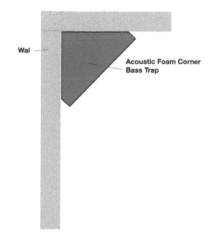

Wal

Acoustic Foam Corner Bass Trap

This type of trap is most effective when occupying the full height or length of a corner.

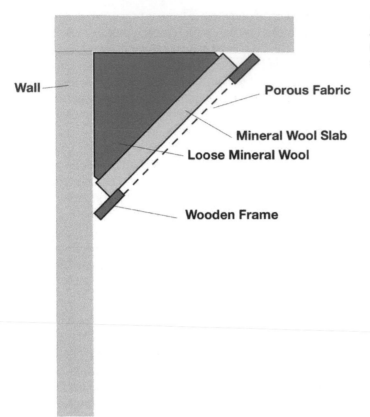

Wall

Porous Fabric

Mineral Wool Slab
Loose Mineral Wool

Wooden Frame

◄

An affordable, DIY bass trap can be created by building a frame to hold a high-density mineral wool slab diagonally across a corner.

original polythene packing. The polythene allows most of the bass energy to pass through so you still get a worthwhile degree of bass absorption.

Limp-mass Absorbers

Limp-mass Absorbers are a form of 'membrane' absorber often constructed from a material known as 'barrier mat' or 'sheet rock'. This is a flexible but incredibly heavy, mineral-loaded material that is physically not unlike very heavy linoleum (but more expensive!). It is typically 10 or 20kg per square metre and the best type has a fabric reinforced backing to prevent it from deforming under its own weight.

The idea of a membrane absorber is that the sound wave energy is absorbed through frictional losses as it tries to move the membrane. Some system use aluminium foil between sheets of mineral wool, others use forms of polythene in a similar way. The materials and the design determine the frequency range affected. Barrier mat absorbs low frequency energy quite efficiently if you hang an area of this material in the path of the sound wave, because as the sound energy attempts to force the sheet into vibration, the heavy self-damping nature of the material absorbs much of the energy through frictional losses. I like to visualise the process as the difference between someone trying to bounce a tennis ball off a solid wall, then trying the same thing against a heavy carpet hanging on a washing line. Because the carpet 'gives' and absorbs some of the energy, the ball doesn't bounce far after hitting the carpet whereas it would bounce a long way after hitting a solid wall. The heavier the ball, the less bounce there would be, and similarly with barrier mat; the lower the frequency, the more effective it becomes.

While limp-mass absorbers are effective at absorbing low frequencies, they still need to be spaced as far from the wall as possible so that they are in the zone where air movement occurs rather than in the pressure zone next to the wall. A distance of 200mm or more is desirable although you can also use this material across corners. The traps also need to cover a large area so that they interact with as much of the low-frequency wavefront as possible – a perfect bass trap of this type can be no more effective than the same area of open window. The nature of the material also means that the surface is also fairly reflective at higher frequencies, so it makes sense to hang them behind foam or mineral-wool absorbers if you want to create a full-range trap.

During the course of our *Studio SOS* series we've used barrier mat 'curtains' hung behind mineral wool-based wall traps, and also behind mineral wool fronted corner traps. We also used it in the construction of a couple of vocal booths to improve low frequency absorption in conjunction with hanging curtains of carpet felt in front of the barrier mat to soak up the highs. It is important to leave space for the barrier mat to hang freely and to secure it along the whole length of the top edge due to its heavy weight. As a rule, the heavier the barrier mat, the more effective it is at low frequencies. This material is also very useful

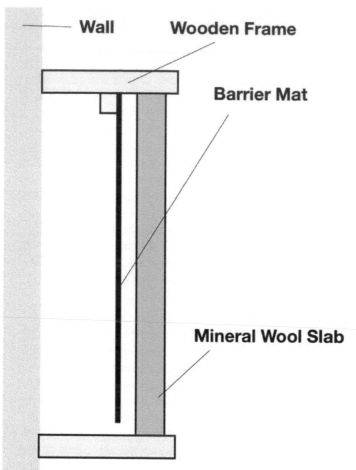

Wall　　　**Wooden Frame**

Barrier Mat

Mineral Wool Slab

◄
You can create a full-range trap by hanging a barrier-mat trap behind a layer of mineral wool, as the low frequencies will pass through the mineral wool layer.

in soundproofing applications where it can be laid directly onto a floor or sandwiched between other building materials such as plasterboard, chipboard and so on, to reduce the transmission of low-frequency sound through the structure.

In small studios where the rear wall is close to the engineer's seat, a good strategy is to try to make the whole of the rear wall, or at least the centre section of it, into a full range trap around 200mm or more deep, and using barrier matting in conjunction with mineral wool, carpet felt, or foam is the most effective and simple way of achieving that.

Free Bass Absorption

Low frequencies are reflected most effectively from solid walls made from dense fixed materials such as stone, brick or concrete. Stud partition or 'dry-walls' made from plasterboard fixed to wooden studding inevitably allow some bass to pass through the wall. However, a lightweight wall also has some 'give' and draws some energy from the sound wave as the sound energy tries to force the material into vibration. This means that home studios set up in rooms with dry-wall construction have an advantage from an acoustic point of view, although they may not be so good from the sound leakage perspective. Dry-wall can also resonate, returning some sound back to the room, so adding a second plasterboard sheet bonded to the first with a resilient glue layer can greatly reduce the problem.

Windows and most domestic doors are also too thin and flexible to reflect much bass energy, so much of it passes though. Obviously, if some sound energy passes through a surface it can't then be reflected back into the room, so although you may face other issues relating to sound isolation and annoyed neighbours, from an acoustics perspective you're better off than someone working in a very solidly built room or in a basement!

Some soft-furnishings and other items of furniture also provides accidental bass trapping, which is why it is so difficult to calculate the acoustic treatment required for small rooms. Fortunately, you can't have too much bass trapping as the overall level of bass put into the room by the loudspeakers isn't significantly reduced by the process. Bass trapping is all about reducing the reflection of low frequencies that would otherwise augment or cancel the direct sound from the loudspeakers. It doesn't take bass out of the room – it stops the bass from bouncing around inside the room in unhelpful ways! The main difference you'll notice when the amount of bass trapping is sufficient is that the bass notes in a piece of music now sound much more consistent in level, whereas in an untrapped or inadequately trapped room some bass notes may sound very loudly whereas others are virtually inaudible.

Commercial Products

In addition to the acoustic foam panels and corner wedges already described, there are numerous commercial traps that combine the benefits of porous and membrane absorbers. These tend to be around 100mm thick and often work best when placed across corners as this improves their ability to trap low frequencies. There are also free-standing panels that can be deployed in the control room or used as movable absorbers in a live room. Often the manufacturers have websites where you can find out how much treatment you need for any given size and shape of room, and some even let you feed in your own room specifications and they'll work out a solution for you. Many of these companies provide lots of additional background information on their websites about the principles of acoustic treatment, making them a valuable free resource.

Chapter Three
Practical Monitoring Solutions

Having devoted a couple of chapters to the basic principles involved, it's now time to see how they can be applied in a real situation. The procedure that follows is more or less the same as we adopt during a typical *Studio SOS* visit.

Before touching anything, the first thing to do is to play some familiar commercial recordings over the monitor system and simply listen. Is the stereo imaging good? Are the speakers wired out of phase? Is the bass end even? We also check out the choice of speakers to ensure they are suitable for the room. In smaller rooms, large speakers with an extended bass response can cause problems that practical bass-trapping solutions can't address in the space available. In such cases the most pragmatic approach is to use smaller speakers with a nominal bass response extending only down to maybe 60 or 70Hz, and to site them as close to the mix position as is practical. In most home studios, a two-way speaker with a woofer diameter between five and eight inches is appropriate.

We also check that the speaker's configuration switches and EQ controls are set correctly for their position (half-space or quarter-space compensation, bass and HF lift and cut, etc.) and that the gain structure is sensible from the interface output through any monitor controller and on to the speaker's input.

In many home studios, the most pragmatic approach is to use small speakers with a limited bass response, and to site them as close to the mix position as is practical.

Gain Structure

Audio systems work best when the audio signal level through the electronics is both well above the inherent noise floor and comfortably below the clipping level, and the concept of 'gain structure' is all about achieving that happy state of affairs. The volume control in the monitor section of a mixing console, computer interface, or on a dedicated monitor controller is designed to work best when used around its nominal '0' position, which is when the knob is pointing around the two o'clock mark on most systems. The volume control provides precise level adjustment around this region, with accurate matching between the left and right channels.

If the gain structure is misadjusted so that the control has to be used at very low settings, the electronics are at risk of

introducing distortion and the sudden development of a fault could easily result in damaging volumes to ears and speakers! It is also likely that the level matching between channels will be very poor, with the stereo image shifting from side to side as the control is adjusted. Conversely, if the level control has to be used flat-out all the time, the system may be noisier than it should be.

To optimise the gain structure, turn the level controls on the loudspeakers to their minimum sensitivity and replay music at a normal level from the computer. Set the volume control on the mixer, interface or monitor controller to its reference position and then adjust the level controls on the two loudspeakers to produce a comfortable listening level, making sure that the two level controls are set to roughly the same position. Finally, switch the music to mono and check that the stereo image collapses to a well-defined narrow image exactly midway between the two speakers. You may need to adjust the input level of one of the loudspeakers slightly to get this precisely right.

A comfortable listening level is a subjective thing, of course. It needs to be loud enough to hear the detail of a recording, but not so loud that you risk hearing damage. If you have to raise your voice to talk to someone close to you, it's too loud! Professional studios often calibrate the reference speaker level to ensure consistent mix balances between different sessions and different studios, and that is generally a good idea. However, the 'industry standard' reference level of 85dB(C) SPL at the listening position (when replaying pink noise with an average (RMS) level of −20dBFS) is often far too loud for a typical home studio.

Room size plays an important part in the perception of volume, and we'd recommend setting a reference level somewhere between 75 and 80dB(C) depending on room size and the amount of acoustic treatment in place. The actual level chosen as 'the reference' isn't critical – what's important is the ability to reset the listening volume to that standard setting when starting or comparing mixes to ensure consistent balances. If you set a standard listening level with the monitor volume control at the nominal '0' position, you can still turn things up briefly to listen to quiet details (or to blow the cobwebs out!) when necessary, but you'll also be able to work at a consistent level when you carry on working on a mix day after day.

If your speakers are connected out of phase, the cones on the monitors will be moving in opposite directions when the same signal is applied to both, causing most of the bass end to vanish.

Speaker Phasing

If the speakers are wired out of phase, the speaker cones on the two monitors will be moving in opposite directions when the same signal is applied to both. Most of the bass end will vanish due to phase cancellation, and if you're sensitive to such things, you'll also experience an uncomfortable itching sensation in your ears as you approach the point exactly midway between the two speakers. It feels a bit like your ears are being sucked out! (Some people find this situation extremely unpleasant, while others barely notice it). If you are using passive loudspeakers with a separate power amplifier, the most common cause is simply an accidental reversal of the two wires connecting one of the speakers, either at the speaker end, or the amp end. With active speakers the cause is usually an incorrectly wired, balanced signal cable.

If the stereo imaging is poor or confused-sounding, with central sources being inadequately defined or blurred,

the usual culprit is something nearby generating strong reflections. A common cause is reflection from computer screens or other hardware placed between and in front of the speakers, but it can also be caused by reflections of the desk or the room's side walls or ceiling. If the room boundaries are causing the problem, poor imaging can usually be fixed fairly easily using properly positioned absorbers. We touched upon that in the last chapter and will be going into more detail later. If the studio isn't set up symmetrically along the centre axis of the room you should endeavour to get your monitors and listening position as close to this ideal as possible before proceeding – even if you need to offset it slightly later on to help smooth out bass problems.

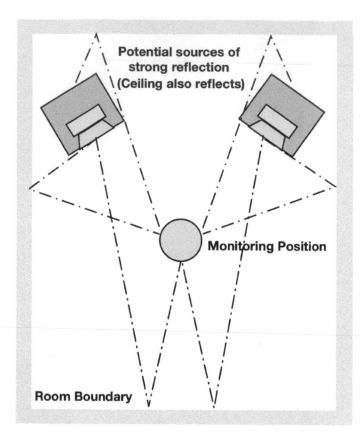

Potential sources of strong reflection (Ceiling also reflects)

Monitoring Position

Room Boundary

◄

Although there will also be reflection from objects in the room, the side and rear walls and the ceiling are always the most significant sources of reflection.

Suitable Monitors

Some speakers simply aren't very good at producing solid, stable stereo images. You should generally pick a speaker designed specifically for studio monitoring, and select one on the basis of accuracy rather than a propensity to flatter the music being played. While flattery may be acceptable and even desirable in a consumer sound system, it doesn't do anybody any favours in the studio. We have come across studios where hi-fi speakers have been used instead of monitors and we've even seen people trying to mix on their band's PA speakers – but neither is a good idea!

Studio monitors are available as either active or passive models – the difference is that passive designs require an external power amplifier while active types have their power amplification built into the cabinet. Both are capable of excellent performance if designed and set up well, but in most cases active speakers are more convenient and negate the need to find a suitable power amplifier. One point to be aware of, though, when using active monitors is that you will need to arrange some means of controlling the overall volume. Analogue or digital mixing consoles invariably have a 'control room' volume control in their monitoring section, but in a mixer-less computer-based system you may not always find a monitor volume control on your audio interface or soundcard. Adjusting the volume in software isn't an ideal solution either since the signal quality can be degraded except when the level is turned up full. There's also no fast way to turn off the sound if your system crashes, outputting white noise to your speakers at digital full-scale level!

The neatest solution is to use a separate analogue monitor controller – a hardware unit, offering level control, source switching, the ability to feed two or more different pairs of monitors, and usually a headphone output. Other useful features might Include a mono button, and a 'dim' button for dropping the speaker level when you want to talk or answer the phone without disturbing your reference listening level setting on the main volume control.

> If you choose active monitors, it is a good idea to also budget for a monitor controller — adjusting the volume in software isn't an ideal solution as it may be impossible to turn off the sound if your system crashes.

If you choose passive monitors it is important to team them with a well-specified power amplifier that has a power rating slightly higher than the upper end of the manufacturer's recommended range for that speaker. Having insufficient power on hand is likely to compromise the sound by running into distortion when you turn up the volume. Whilst distortion is clearly a bad thing for accurate monitoring, it can also damage tweeters extremely quickly, so in most cases having a little excess power available is far better than not having enough.

Speaker Positioning

After our initial listening tests we check the speaker mounting arrangements and also their alignment. In general, the two speakers and the listening position should all form the corners of an equilateral triangle, with the tweeters aimed towards the engineer's head in both the vertical and horizontal planes. However, it is worth checking the speaker manufacturer's own recommendations, as some designs are intended to be aimed slightly behind the listening position, or even to point straight down the room.

The moving cones within a speaker don't just create sound, they also cause the cabinet to move and vibrate in a way that can degrade the sound or cause other things to vibrate in sympathy, so it is very important that these unwanted vibrations are controlled. If the speakers are mounted on floor stands, they should be fixed solidly to the stand using something like Blu-tak, and the stands properly levelled and made secure on the floor, using spikes to penetrate any carpet present. However, if the speakers are to be mounted on a desk or shelf we have found it is generally much better to use foam isolating pads, as these keep the speakers stable but prevent vibrations reaching the desk where they might set off sympathetic resonances or other unwanted rattles and vibrations. We also check that speakers with rear bass ports or passive radiators are not positioned closer than six inches from a wall (and ideally more).

We mentioned earlier that foam pads can work better with a thick ceramic floor tile or metal plate fixed to the top. Having a solid platform reduces the amount of energy transmitted through the foam but more importantly it adds weight and stability to the speaker cabinet. Whenever a speaker cone moves forwards, the cabinet tries to move in the opposite direction, according to Newton's laws of motion, so adding mass and damping helps limit this movement. If the cabinet moves in a reaction to each bass note, it takes the tweeter with it, which introduces a form of modulation that can colour the sound. The less the cabinet moves, the less the sonic colouration and the tighter the sound generally. Commercial platforms made with steel tops often have an integrated non-slip rubber mat on top but you can use non-slip kitchen matting on your DIY floor tile versions. This matting also works well on speaker stands as an alternative to Blu-tak.

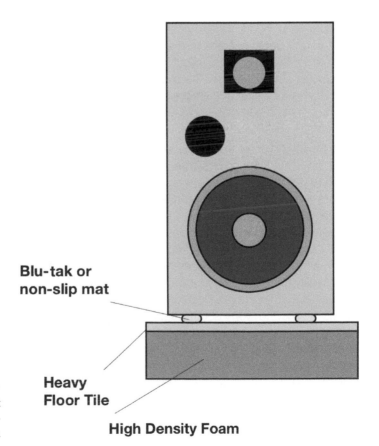

**Blu-tak or
non-slip mat**

**Heavy
Floor Tile**

High Density Foam

A DIY speaker platform – the Blu-tak ensures that the speaker does not slip on the surface of the hi-mass layer.

Uneven Bass

If the bass response sounds uneven, with some boomy notes and some sounding very weak, the usual cause is unhelpful room modes, but improvements can often also be made just by repositioning the speakers. Moving them forward or back, or side-to-side, a few inches can make a surprisingly big difference. The trick is to move the speakers, run the chromatic sine-wave test outlined in Chapter One, then see if the result is better or worse before making further moves. This technique has proven to be a life-saver on many *Studio SOS* jobs where all other attempts to create a more even bass response have failed.

We believe that you'll invariably get the best results by having the speakers fire down the length of the room unless the room is at least 15 feet wide, and during the course of our *Studio SOS* visits we've had to disassemble more than one studio and turn it through ninety degrees to enable us to get acceptable results. In some cases a bad layout had been chosen by the studio owner simply because that's as far as their cables would reach! Buying (or better still, making) a few new cables is not a big price to pay to get your studio working at its best.

Where the studio is setup in a typical domestic room constructed with lightweight plasterboard walls, you should be able to get a reasonably even low end from modestly sized monitors without having to resort to installing much, if any, bass trapping, as this type of structure naturally absorbs some low frequencies and lets a lot straight through. However, if trapping is required then corner traps are the most effective solution where space is limited, as we saw in Chapter Two, and it is fairly easy to make your own if you don't want to buy commercial foam corner wedges or ready-made panel traps. There is no need to worry about installing too much bass trapping because you won't accidentally kill the low end! Bass trapping doesn't affect the amount of bass coming from your loudspeakers, it just stops that bass from reflecting around the room and interacting with the direct sound, which is what causes the unwanted peaks and nulls at different frequencies. So bass trapping doesn't reduce the bass, but rather your bass end becomes much more even and consistent.

Subwoofers

We often get asked if a subwoofer is appropriate in a small studio. Our general view is that a sub that generates very deep low frequencies will often just emphasise the modal problems caused by room geometry, and so often does more harm than good. Also, cheap subwoofers produce a lot of harmonic distortion that masks the midrange detail of the main monitors, so they can easily make a system sound worse even at higher frequencies! However, so-called 2.1 systems that combine very small main monitors with a dedicated subwoofer are increasingly popular and do have some advantages. Amongst these is that in a very small room you may not be able to reposition conventional speakers to achieve the flattest bass response without severely compromising other aspects of the setup. However, with a 2.1 system you may be able to place the stereo speakers to

One way of finding the best location for a sub is to place it on the floor where you would normally be sitting, and then crawl around the floor, so your ears are at the same height as the sub, until you find the position where the bass response is most even – that's where you should put the sub!

achieve good stereo imaging, away from walls and corners, while also positioning the separate subwoofer somewhere that gives the most even bass response.

Unfortunately, people often place a subwoofer for convenience rather than where it actually works best. We've found subwoofers to be very position-sensitive and, as a rule, they definitely shouldn't be positioned exactly midway between two walls as that excites the room modes more than if the sub is slightly off centre. Even though bass is largely non-directional, the sub should still be placed in front of the listener because the inherent harmonic distortion will generate higher frequencies that can be located, and the combination can start to sound disembodied if the subwoofer is placed off to the side or behind the listening position. We've come across situations where the user has put the sub on a partly enclosed shelf, in a makeshift box, or in the centre of the room – and in all cases we had to move it to a more appropriate position to achieve a more uniform bass response. Subs shouldn't be placed in confined spaces that are likely to resonate, and more often than not they work best sitting on the floor somewhere close to the front wall, though if this isn't physically possible you'll probably find an acceptable spot towards the front of one of the side walls.

One commonly employed technique for finding the best location for a sub is to first place the subwoofer on the floor right where the engineer's chair normally sits, and then to play some well recorded music with a lot of different bass notes (the bass-note test sequence described in Chapter One would work well, too, with its playback looped). You then crawl around the edges of the room with your ears at subwoofer height, listening for a spot where the bass sounds most even. Once a suitable location has been found, move the subwoofer there and check that the response is still even when you are at the usual listening position. You can then start the process of optimising the subwoofer's level, phase and cut-off frequency to blend its contribution into the main satellite speakers. This positioning technique invariably gets good results and we have used it successfully many times. The idea is simple – to swap the positions of the listener and the sub while searching for the most even-sounding spot – and it works!

Of course, it's very easy to say 'optimise the subwoofer's level, phase and cut-off frequency' but a very common problem we find is that users often turn the subwoofer level up much too high. The idea of the sub is to underpin those low octaves that the main satellite speakers can't quite manage, so its contribution should only be noticed by its absence when you switch it off! It might be good fun to have your trousers flapping in the breeze and your internal organs rearranged by excessive amounts of low-frequency energy, but it won't do your mixes any favours! If in doubt, set the subwoofer level too low rather than too high, and if the manufacturer offers a calibration procedure, follow it!

/ SIMPLE SUBWOOFER CALIBRATION

A simple calibration procedure that works with most subwoofers is to first set its phase, then the cut-off frequency, and finally the level. Not all sub-woofers have all three controls, so some experimentation and compromise is often required. Since the subwoofer is unlikely to be placed under the satel-lite speakers its sound contribution may have further to travel to the listen-ing position, and usually a phase control is provided to compensate for that. The easiest way to find the optimum setting is to play a sine wave note at a frequency that both the satellite speaker and subwoofer can reproduce – typically somewhere between 80Hz (E2) and 120Hz (B2) – and then adjust the phase control to find the position where the sound is loudest – which indicates that the contributions from the sub and satellites are all arriving in phase at the listening position.

Next, using the chromatic sine wave sequence again playing just over the satellite speakers, identify the few notes below which the level starts to fall off significantly. Then, playing the same sequence over the subwoofer, adjust the cut-off frequency control until the level falls off rapidly above those same notes. Don't get bogged down over this too much – it's only a rough setting at this stage.

Finally, playing some full-range music over the complete sat/sub monitoring system, adjust the level control so that the subwoofer is just reinforcing the lowest notes. At this stage you may need to fine-tune the cut-off frequency and level controls slightly – they will tend to interact with each other – to optimise the integration of the subwoofer with the satellites. The aim is to be completely unable to tell when the satellites hand over to the subwoofer, and to only notice what the subwoofer is doing when you switch it off!

Set phase first, then adjust cutoff frequency and level – you may also have to experiment with different placements.

Fixing the Imaging

Chapter Two explained the role of acoustic absorbers and how they can benefit your room acoustics, so the next stage is to figure out the best places to put them. Before doing this though, it is important to appreciate that you can easily over-damp a room, which not only makes it sound oppressive but also disturbs the even decay time we've been trying to achieve. As a very general rule, we'd recommend not covering more than around 20% of the available wall space with mid/high absorbers. Some of the studios of the 1970s with mineral wool covering all the walls sounded dreadful. That's because if you apply too much broadband absorption, the room will start to sound coloured due to the inescapable fact that simple absorbers don't absorb all frequencies equally, typically offering very little absorption at bass frequencies. Where mid and high frequencies are concerned, our primary aim is to achieve a listening position free from strong first reflections.

Mirror Points

The most important areas to treat are the so-called mirror points – those areas of reflective surface that bounce the sound from the monitors directly back to your listening position. You can find these by getting someone to hold a real mirror flat against the wall when you are seated in your usual monitoring position, as described in Chapter One. At any point where you can see a reflection of either monitor speaker in the mirror, you need to place an absorber. The main mirror points in a typical rectangular room will be on either side of you, slightly forward of where you sit, and also on the wall behind the speakers. In a typical bedroom or garage studio, you may find that one square metre of suitably positioned absorber per wall is enough. In general use the thickest acoustic absorbers you can accommodate, and ideally leave air gaps behind the foam if at all possible as this makes the traps more efficient at lower frequencies. You may remember that we suggested that if mounting foam on board, then select a board with generous perforations where possible because with solid MDF or plywood boards any benefits of leaving an air gap behind the trap are lost.

There will also be an area on the ceiling, roughly halfway between your monitors and your ears, where an absorber could be helpful. If you don't want to stick foam to the ceiling, build a lightweight wooden frame to hold a sheet of foam or mineral wool and suspend it on wires or chains from cup hooks or plasterboard fittings screwed into the ceiling, which is far less intrusive than using glue. Hang your absorber roughly

Any non-absorbent surface on which you can place a mirror and see your speakers reflected is a potential source of reflected sound.

100 to 200mm below the ceiling to increase its low-frequency efficiency. If there are lights above the ceiling panel, consider fitting LED lamps to avoid heat build-up.

Commercial acoustic panels are available in both square (typically 600x600mm) and rectangular (600x1200mm) forms, and we're often asked whether it's best to hang rectangular acoustic panels vertically or horizontally. The answer is that it depends on the room size and layout as well as the aesthetics – there is no right or wrong way! In larger rooms, you can also double up on the panel area.

We've come across many situations where the ideal side-wall absorber placement has coincided with entrance doors, cupboard doors or windows. One solution is to make up a removable lightweight foam absorber that hangs from a hook on the back of the door or cupboard. Self-adhesive plastic

A room with all the primary reflection points treated.

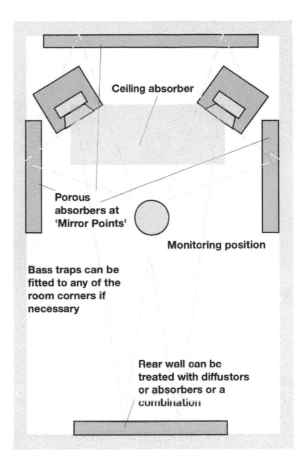

Ceiling absorber

Porous absorbers at 'Mirror Points'

Monitoring position

Bass traps can be fitted to any of the room corners if necessary

Rear wall can be treated with diffustors or absorbers or a combination

hooks are usually fine for this and you can stick an old CD onto the back of the foam panel to hang over the hook. Acoustic foam can be fixed using the recommended spray contact adhesive, but be warned that once fixed to a hard surface it puts up a real fight if you try to remove it at a later date, and you'll probably end up with a layer of foam still stuck to the wall! Often it is safer to fix the foam to an MDF or plywood panel, or easier still an old CD, that you then hang on a picture hook. Spray adhesive designed for acoustic foam is the best option, though regular contact adhesive can be used for fixing small items, such as CDs, to the back of a panel. Better grades of carpet adhesive also work but we've found that some brands don't grip strongly enough, and some can dissolve the foam – so always try a small area first!

▲
Gluing an unwanted CD to the back of a foam tile allows you to hang it on a picture hook so that it can be placed temporarily and removed without damaging the wall.

If a mirror point coincides with a window, then heavy curtains are the simplest answer, although we've also had good results making a removable foam absorber panel which could be hung up or balanced on the window ledge only when mixing. Vertical blinds set at a 45-degree angle can also be useful in breaking up reflections. Movable panels are useful in small studios anyway as during tracking you can use them to improve the acoustics around an instrument or amplifier when recording with microphones.

In some rooms perfect symmetry simply isn't possible, as we've sometimes found in bedroom studios we've visited. All you can do in such cases is to try to make the listening area as symmetrical as possible from a mid- and high-frequency point of view, and this can often be achieved by hanging a temporary foam panel to one side or other of the monitoring position when mixing. For example, if your desk is close to a wall on your left, then prop up a temporary foam panel on your right to make something resembling another similarly treated wall. Another easy temporary fix is to hang a foam panel wherever necessary from a boom mic stand using a couple of those woodworking clamps that look like giant clothes pins.

TIP: If you accidentally get some spray adhesive on the surface of acoustic foam or other furnishings, you can often remove it before it dries by dabbing at it with the sticky side of Gaffa or duct tape.

If you can't permanently mount any sound absorption in your studio room, you can always improvise some temporary absorbers when you are actually working.

Rear Walls

In very small rooms you may also need to place absorbers on the wall behind you, but how best to treat the rear wall depends largely on the size of the room. The basic idea is to prevent strong reflections from the back wall bouncing back to the listening position.

In a commercial studio the room is often large enough that diffusers can be used to scatter sound that hits the rear wall, which has the advantage of keeping the sound energy in the room (helping to prevent it from sounding oppressively dead), but without causing any strong reflections. However, where the rear wall is closer than about six feet from the back of your listening chair there is insufficient space for pure diffusion to work effectively, so using absorbers or a combination of scattering and absorbers, offers the best practical solution.

Although a little more costly to build, a full-range absorber is the most effective treatment, which you can create by hanging barrier mat 200mm or so from the wall and then putting a layer of foam or mineral wool in front, making sure that the barrier mat hangs freely. If you're proficient at DIY, you can build a simple wooden frame to accommodate the trap and then finish off the front using acoustic foam or fabric. However, make sure that the frame includes plenty of strong cross-members as experience has shown that studio visitors invariably try to lean on fabric-covered traps, which can do a lot of damage!

The cheapest, simplest and most pragmatic solution for small and medium-sized rooms is often to combine absorption with some crude scattering diffusion provided by shelves full of irregularly sized books, CDs, software boxes and unused equipment. The ubiquitous studio sofa (or even a bed and mattress) makes a reasonably helpful absorber (ideally covered in fabric rather than leather or vinyl, as the latter reflect high frequencies), and you may even be able to stuff its underside with mineral wool to make it more effective as a bass trap. Exposed parts of the rear wall may then be treated with areas of acoustic foam or other DIY absorbers.

<
The most pragmatic solution for the rear wall of many small home-studio rooms is simply to rely on the degree of scattering diffusion provided by shelves full of irregularly sized books, CDs, software boxes and unused equipment.

Of course, you have to be realistic about what can be achieved in a small project studio and often some acoustic problems remain that you simply have to work around. A case in point is the cube-shaped room where nothing you do is likely make the centre of the room usable as a monitoring position, especially if the walls are solid rather than drywall, because by the time you have installed enough bass trapping to tackle the problem you probably wouldn't be able to get into the room anymore! So, even if you sit in the middle while tracking you will need to push your chair out of that central 'dead zone' to make meaningful mix decisions, and working with headphones might actually be the better solution.

Checking your mixes on headphones is a good idea, even in a well-treated room.

Having said that, the basic room treatment we've described can still make a hugely noticeably improvement to both the overall focus and stereo imaging of the sound, and if you keep comparing your mix-in-progress with commercial tracks in a similar style played over the same system you'll almost certainly be able to produce good results. Double-checking your mixes on headphones is also good practice. It can be a good idea no matter what kind of monitoring you have, but it is especially beneficial if you know your usual monitoring environment is not entirely trustworthy.

TIP: If you know that your monitoring is inaccurate at low frequencies, avoid applying EQ to frequencies below 80Hz or thereabouts unless you can check the results on headphones and other people's sound systems. If you're using samples for the drums and bass instruments, it is safer to trust that the sample designers have got the bass balance about right than to attempt EQ adjustment in a poor monitoring environment. The danger is that you'll inadvertently be equalising to compensate for room response anomalies, rather than genuine deficiencies of the track!

/ MONITORING DO'S AND DONT'S

DO arrange your monitoring setup as symmetrically as possible.

DO put your speakers on rigid stands or, if mounted on a desk, isolating pads.

DO place acoustic absorbers at the mirror points.

DO move your monitor position to achieve the most even bass.

DO scatter or absorb reflections from the wall behind you.

DO continually compare your mixes with commercial mixes in a similar style.

DON'T allow objects to get between you and the monitor speakers.

DON'T site monitors too close to room corners or back walls.

DON'T sit in the exact centre of a square or rectangular room.

DON'T sit close to walls or corners when monitoring.

DON'T use too much broadband absorption or the sound of the room may suffer.

DON'T rely too much on your monitors in a small room – double-check the mix on headphones and on as many third-party sound systems as possible.

/

Chapter Four
The Recording Space

The majority of project studio operators will have a requirement to record vocals at some time or other, even if all their other instruments come from samplers, soft synths and commercial loops. As long as there's not too much sound leakage from the outside world or from excessively noisy computers, it's actually pretty easy to make good vocal recordings almost anywhere, but there's more to it than simply buying a good mic and preamp. The acoustics of the recording space are massively more important than the precise make and model of microphone or preamp, yet they're all too often not given the same degree of consideration. Fortunately, when you're recording vocals you don't have to treat the whole space, only the area around the microphone and singer.

You may have seen adverts for special acoustic screens, designed to be positioned directly behind a microphone, and we often use them as part of our *Studio SOS* solutions. However, while these devices can certainly help to reduce the amount of reflected room sound getting into the rear and sides of the mic, they don't offer any protection from sound coming from above, or indeed sound reflected from the wall behind the singer. The latter is usually the most significant source of unwanted 'room sound' in a vocal recording, because that's the direction in which the microphone is most sensitive.

For this reason we always recommend that such screens be used in combination with a large absorber hung directly behind (and if possible around the sides of) the singer. Where the ceiling is low, a piece of acoustic foam suspended above the singer and mic often helps as well. We always raise a smile when mentioning duvets as suitable absorbers for vocal

The acoustics of the recording space are massively more important than the precise make and model of microphone or preamp, yet they're all too often not given the same degree of consideration.

recording, but they're cheap, they're readily available, and they work remarkably well.

Our typical setup uses a vocal acoustic screen placed so that the mic is roughly level with the front of it, and the 'live' side of the mic pointing outwards! If the mic is placed too far inside a curved screen internal reflections can cause obvious colourations. The duvet (or other appropriate absorber) works best if placed across the corner of a room and arranged to hang directly behind the singer. Where necessary, a single 600 x 600mm square of acoustic foam is suspended above the singer by whatever means we can improvise at the time – on another mic stand, or from cord fixed to pins or hooks in the ceiling, for example. Most people are surprised at the dramatic

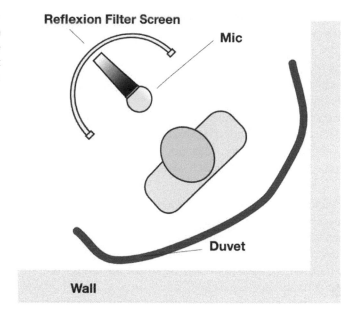

Reflexion Filter Screen

Mic

Duvet

Wall

improvement in vocal sound these simple measures provide, even though they may have read about us doing exactly this on many occasions. Without this basic acoustic treatment a vocal recording will always be compromised by room reflections and, once compression has been used to process the track, the vocal will sound very boxy. No amount of processing can rescue a poor vocal recording with a lot of boxy room colouration. You can read more about this approach in Chapter Seven where we look at vocal recording in more depth.

Vocal Booths

Improvised duvet absorbers are cheap and reliable, but are only suitable where there is little background noise to worry about. If ambient noise is an issue then a proper vocal booth may be the only answer, although we usually try to talk people out of building stand-alone vocal booths if we can! It is all too easy to end up with something that sounds boxy and horrible – indeed we've had to try to rescue a few poorly designed vocal booths. The usual problem is a vocal booth which is far too small (and so inevitably sounds boxy), which has been treated with an excessive amount of mid and HF absorption to try to remove the boxiness, leaving it sounding very dead but also boomy and bass-heavy.

The more reasonably sized but over-damped booths sometimes respond well to adding a little 'hard surface' high frequency reflection around head height using old CDs as reflectors. However, those set up in small cupboards under the stairs are usually irredeemable and we have to persuade the owners that recording vocals out in the main room using the 'screen and duvet' approach will actually give far better results.

On another occasion we found a vocal booth that was adequately large but the acoustic treatment applied inside it was too thin so all the life was sucked out of the high end while leaving the lower mids to run rampant. All that was available to us for impromptu corrective treatment was carpet and rugs, and although any significant area of carpet stuck onto a wall is usually disastrous from an acoustic standpoint (as explained in Chapter Two), we managed to hang some rugs inside the booth spaced away from the walls. This improved the sound to a useful extent because the rugs were behaving a little like our suspended barrier mat low frequency absorbers but with the carpet fabric also soaking up higher frequencies too. Hardly a calculated solution, but it does show how you can often improvise a reasonable solution to a difficult acoustic problem through science and lateral thinking!

For one Studio SOS visit to a home studio in the USA, the owner was using a large walk-in clothes closet as a temporary vocal booth, with long mic and headphone cables running across the bedroom and landing back to his studio room. As the closet was directly behind the back wall of his studio room, we drilled a hole in the studio wall to connect cables through, which was a much more practical and convenient solution. Whenever we've had to make a cable hole through a drywall partition, we've made holes on both sides and then inserted a section of plastic waste pipe to ensure an unobstructed passage for cables – it only needs to be large enough to feed XLR connectors through one at a time. The pipe is then fixed into the wall using flexible mastic filler leaving a nice tidy job. Foam offcuts can be jammed into the pipe to form an effective sound seal.

In this particular example, the closet was around 1700mm square and easily large enough to provide plenty of space for a standing vocalist, while the racks of clothing hanging on three sides provided excellent broadband absorption. Acoustic foam on the ceiling would have made it even better but there's a limit to what you can do in a space designed for living rather than music.

Vocal Booth Treatment

On two *Studio SOS* visits we've helped build vocal booths and, as always, we've learned something from each one. The first was built into a space that had already been constructed by the reader but he left us to handle the internal acoustic treatment. We'd warned him beforehand to make sure the basic room was large enough and it worked out OK.

Our approach was to make the wall behind the singer a full-size trap comprising mineral-wool slabs against the back wall, with a curtain of carpet felt underlay hanging in front of it, and then a sheet of barrier mat, both hanging freely. The idea behind this simple rear wall trap is that the felt absorbs the higher frequencies while the barrier mat sucks the energy out of the lower frequencies. We used a cotton/polyester bed sheet stapled over a wooden frame to hide the rear wall trap, using enough internal bracing to prevent singers leaning into the trap and damaging it.

We also covered the side walls at head height with acoustic foam around 75mm thick and spaced from the wall slightly on foam blocks, but left the lower side walls untreated so as not to kill all the HF reflections. Some foam was also fixed to the ceiling but again we were careful to leave some reflective areas so the high end didn't suffer. The double-glazed entry door was of course entirely reflective though there was room within the booth to fit a curved reflection screen behind the mic. The combined acoustic treatment worked exceptionally well and produced well-balanced, sufficiently dry-sounding vocals.

We also floated the idea of fitting a movable mic arm mounted to a wall or to the ceiling, as this would avoid having to take a mic stand into the booth to clutter up the limited floor space. The owner found one with an arm much like the mechanism of an Anglepoise lamp and it worked out really well. Though not strictly an acoustic consideration, we also suggested that LED lights would cause less heat build-up inside the booth than conventional halogen spots, and these were also installed in due course.

When building our next 'from scratch' booth, we were told we had a space of just over 1220mm square to work with, although we managed to negotiate a little more depth as we wanted to incorporate a similar rear-wall trap that had been so successful previously. The studio owner phoned around all the local double-glazing firms to see if they had a surplus window

The most successful DIY vocal booths incorporate a balance of absorbent and reflective surfaces in their acoustic treatment – it is important to retain some high frequencies to prevent the space sounding boxy and coloured.

(they often have a few windows that were made up to the wrong size or from cancelled orders that they're happy to sell off cheaply) and eventually located a small window that would allow the singer to view any musicians working off to one side of the booth. He also bought a door with double-glazed plain glass in the top half for use as the entry door.

We built the frame of the vocal booth using 50 x 100mm reclaimed timber and then screwed two layers of 12mm plasterboard on either side with loft insulation stuffed into the frame spaces to give us some internal damping. All the joints were sealed with mastic and adhesive was applied to the frame and blobbed between the plasterboard sheets. We decided to cover the outside with thin cord carpet, as that was a cheap way to make the booth look tidy and the carpet also helped damp the panels a little more.

With the right materials and an understanding of the principles involved, you can create a quite sophisticated vocal booth using just basic DIY skills.

1. Installing the frame for the rear-wall limp-mass absorber.

2. Attaching mineral- wool slabs to the rear wall.

3. A layer of heavy felt underlay hangs in front of the back wall mineral wool slabs, with the mineral-loaded vinyl 'barrier mat' sheet hanging in front of that.

4. (Left) Attaching the heavy felt underlay. 5. (Right) Gluing a foam absorber panel to the ceiling.

Combining traditional foam absorbers with a broadband rear-wall trap seemed to be the way to go, again leaving some reflective space around and below the foam. The untreated plasterboard plus the door and window added sufficient reflection to keep the sound reasonably bright, while our hanging absorber and foam panels cleaned up the sound very nicely and balanced the reverberation spectrum.

Instrument Recording Space

There are countless documented examples of how a studio should be designed to accommodate the recording of acoustic instruments, including drum kits, but in the real world of the home/project studio you generally have

You can fabricate simple DIY acoustic screens to achieve more separation between musicians playing at the same time, or even just improvise by hanging duvets, folded blankets or sleeping bags over clothes-drying frames or spare mic stands.

to make the best of what you have, augmented by a little improvisation. Second-hand office dividers form a useful basis for constructing acoustic screens to provide separation between adjacent instruments. However, the thin foam covering they come with is only effective at speech frequencies. One cheap solution to upgrade their performance is just to hang a winter-grade polyester duvet or a few layers of blanket over them, although better solutions involve attaching mineral wool or foam absorption panels, as we described in Chapter Two. You can also improvise acoustic screens simply by hanging duvets, folded blankets or sleeping bags over clothes-drying frames, chair backs or spare mic stands, if you need to dry up the room's acoustics and improve the separation between musicians playing at the same time. In combination with careful microphone placement, such simple measures can make all the difference to the quality of a recording.

Reflections and Resonances

Reflective floors usually help in the recording of acoustic instruments, because most instruments are designed so sound best when played in that kind of environment. So if your studio room has a fitted carpet, laying a sheet of MDF, hardboard or plywood beneath the recording area, extending from the instrument to the microphone, will often help breathe a little life into the high end. We've even resorted to using readers' melamine-covered tea trays and place mats to create a temporary reflective floor area when nothing else was available – and it really worked!

One often overlooked aspect of recording acoustic instruments is to try different parts of the room to find a 'sweet spot'. In Chapter Two where we discussed acoustic treatment we explained how room modes, which are essentially resonances, can affect the bass balance of your monitoring. These same room modes also affect the way an instrument sounds in a room, and what the microphone captures. On one of our *studio SOS* visits, in which we were concentrating on acoustic guitar recording, our host was very surprised at just how dramatically the sound changed when using the same mic setup in different parts of the room.

Drums can pose a particularly difficult recording challenge in smaller rooms, as the detrimental effect of reflections from nearby walls and low ceilings is compounded when multiple mics are used. Close mics on individual drums can usually be positioned to give an acceptably dry, tight sound as the ratio of direct to reflected sound is fairly high – although you may still need to experiment with the kit position within the room to get the best kick and snare sounds because of those pesky room modes again! The overhead mics are more problematic, however, as they generally need to be positioned a metre or more above the height of the tom-toms which, in a domestic room, puts them so close to the ceiling that the strong reflections will cause significant colouration. We will look at a few possible workarounds in Chapter Ten, dedicated to drum recording, but most practical solutions involve placing absorbers between the overhead mics and those surfaces causing reflection problems.

▲

If your studio room is carpeted, laying a sheet of MDF, hardboard or plywood beneath the recording area, extending from the instrument to the microphone, will often help breathe a little life into the high end.

Other Rooms and Spaces

Where the studio room itself is too small to allow much in the way of mic placement experimentation, we have had occasion to try recording instruments in different spaces throughout the owner's house. This can be tricky if you're working entirely alone, although a wireless keyboard can be useful for starting and stopping DAW recorders.

Although your vocals might sound more flattering to you when you sing in a tiled bathroom, the excessive reverberation is rarely appropriate to a contemporary music mix, but the same space might turn out to be well-suited to acoustic guitars, guitar amps, violins, wind instruments, hand percussion and so on, so it's well worth getting yourself long microphone and headphone extension cables to enable you to audition bathrooms, stairwells, hallways and conservatories. Facing a guitar cab down a long corridor, for example, opens up the opportunity to try combined close and distant mic setups that may be physically impossible in a small bedroom studio. Communication with musicians playing elsewhere in the house shouldn't be a problem as you can hear the performer via their instrument mic, and you can talk to them via a talkback mic routed to their headphones.

◄

The tight, bright-sounding ambience of a reflective corridor or bathroom is well suited to acoustic guitars and some electric guitar sounds.

Where additional rooms are off-limits, we've resorted to putting loud guitar speakers or combos in a clothes closet along with the microphone, or building a tent over them using a clothes-drying frame and then heaping blankets and duvets on top – don't do this with a tube combo, however, as it will get very hot! If your tube combo's speaker is not hardwired to the amp, you could unplug it and use the speaker output to feed a separate cabinet to allow you to take advantage of these techniques. Similarly, we've put them behind sofas, facing into the back of the sofa with a duvet or two draped over the top to contain the sound. Because the duvets and blankets are fairly good sound absorbers, containing a guitar speaker in this way doesn't usually make it sound too boxy, and it certainly removes any ambient room sound! More guitar recording approaches are examined in Chapters Eight and Nine.

Keyboards – Real and Virtual

We've written most of this book on the assumption that the reader is working with one of the mainstream DAW programmes, and therefore also utilising its bundled virtual instruments for any keyboard parts that are required. Many of these now offer superb audio quality and playability, and for most people who aren't serious keyboard players, there is little reason to look elsewhere for keyboard sounds. Triggered via MIDI, either down a MIDI cable, or, increasingly, via MIDI over USB, virtual instruments generate their sounds in real-time within the DAW itself, and therefore there are no issues with capturing their audio. You may wish (or need) to bounce a real-time audio source into an actual audio track on occasions, in order to free up some processing resources, but even this presents no challenge in terms of creating the audio recording.

There may be performance issues, caused by a slow response (latency) if your DAW's audio buffer setting is high, but most of the mainstream programmes now have a 'low latency' mode that temporarily disables processor-hungry plug-ins in favour of a faster response to help deal with this situation. The only other factor to think about with virtual instruments is their default output level, which is generally very high. If you want to

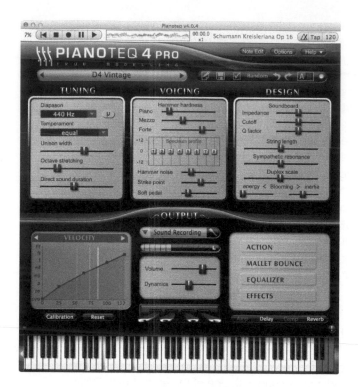

◄

Unless you already have a collection of classic keyboards and the ability to record them well, you'll struggle to create better results than those offered by the best software instrument plug-ins, or even the bundled virtual instruments that come with most of the popular DAW programmes.

leave enough headroom in your mix bus and have your virtual instruments balance well with audio tracks recorded at sensible levels, it is worth cultivating the habit of routinely trimming the virtual instruments' output level settings by 6 to 10dB as soon as you load them up, and saving attenuated versions of your favourite presets.

Recording hardware electric keyboards presents much the same challenge as recording a bass guitar, in so far as a straight, clean DI is often sufficient, but there will also often be instances when you also want the character of an amp/speaker combination as well, or instead.

With a DI'd signal you only really have to be concerned with level matching: if your keyboard requires a power supply, it will almost certainly have a low-impedance, line-level output. Set its output level control to about 75% and there should be no problems at all connecting to a dedicated line input or, by using

DI boxes provide level and impedance matching, but they also offer isolation of the source and destination ground connections, thereby minimizing the risk of ground-loop hum getting into the signal path.

a passive DI box, to a microphone input. Using a DI box has the advantage of allowing the input and output ground paths to be kept separate to avoid the possibility of a ground-loop hum problem.

If you are using an active DI box to feed the signal to a microphone input, however, you should look for the input attenuation switch and drop it to at least its first attenuation setting. The unattenuated setting on an active DI box is usually optimised for a passive instrument connection and may well be overloaded by the line-level output of an active keyboard instrument. Recording a line-level source is easy, but you can still get it wrong if you fail to understand the gain structure.

Using keyboards with amps requires recording via mics, but as you will probably only be doing this in the pursuit of 'character' in the sound, there really is no right or wrong. Dynamic mics will offer plenty of midrange punch, especially on electric piano, whilst ribbon and capacitor models will sound a little more open and balanced. The techniques are very little different than recording guitar amps, benefiting equally from treatment of the primary reflection areas and control of any major room colouration.

Acoustic piano recording is in many ways analogous to recording acoustic guitars, as discussed in Chapter Eight: the choices between detail or distance, and true stereo or spread spot mics, are actually precisely the same. Small-diaphragm capacitor mics are often preferred in this application, for their low off-axis colouration – piano presents a wide sound source, even when close-miked. The further back your mics are, the more of the room they will be 'hearing' and piano is such a familiar sound to most of us that excessive room colouration is particularly noticeable in piano recordings.

Treatment of the environment in all the ways described in this book will make a worthwhile improvement to the sound of a good piano in a poor room, but the pragmatic solution in the majority of home studios lucky enough to have an acoustic piano, is to close mic the frame about 12 inches back from the hammers with a pair of cardioid or omnidirectional microphones, and then create an artificial environment with a good-quality digital reverb.

Chapter Five
Soundproofing

A common complaint from project-studio operators is that too much sound either leaks into or out of their studios. Clearly, sound leaking out can cause problems with neighbours or other members of the household, while sound leaking in may get onto recordings made using microphones. However, the techniques used to stop sound getting in will also prevent from it getting out, so whether your concern is leakage in or leakage out, the approach to fixing it will be the same.

The subject of soundproofing is very different from that of room acoustics and is generally more complex to implement in a project studio situation, but sometimes attention to this area is necessary to make the operation of the studio viable. Unfortunately there is no lightweight solution to problems in this area, and foam and fibre materials that work to improve the acoustics of your room do little or nothing to improve sound isolation.

At the outset maybe we should drop the term soundproofing and replace it with the term 'sound isolation' as there is rarely such thing as total soundproofing, at least when talking about domestic studios. In reality the best you can hope for is to improve the situation, so you really need to determine whether the measures that you are able to take (both practically and economically) can reduce sound leakage to an acceptable level before you spend any money. For example, if you live in a rented apartment with only drywall partitions between you and the neighbours, it would be wishful thinking to imagine that there could be any practical or affordable solution that would allow you to practise on your double-kick drum, power-metal kit at 2am without everyone else in the building being aware of it. One reader contacted us to ask about a cheap soundproofing solution because they were only playing cheap instruments. Sadly the laws of physics don't care about your budget! On the other hand, if you have more realistic expectations there's a lot that can be done without breaking the bank, and a little DIY skill can save you a lot of money. Most of the isolation improvement solutions outlined in this chapter are within the scope of a DIY enthusiast, mostly using common materials available from your local builders' merchant, although some more specialised items may be required from a studio materials supplier.

The quest for sound isolation has many different paths, one of the more common ones in professional circles being the construction of a full 'room within a room' – a completely separate inner room inside the existing space and isolated from the original floor by blocks of neoprene rubber or suspended on metal springs. This, however, is hardly a suitable approach for most home studios where space is often already limited, although if you're setting up in a rented industrial unit it may be worth considering. There's plenty of documentation on the subject available in acoustic design reference books and online, but it is rather beyond the scope of most DIY studio projects. Fortunately, you rarely need to go quite that far – we've seen a number of perfectly adequate studios, often in colleges, where additional studding and plasterboard walls have simply been added to the existing structure to give a worthwhile improvement.

Where Does Sound Go?

As with any other form of energy, sound energy can't be destroyed or lost, it can only be converted to another form of energy. The vibrational energy of sound is converted into heat via the mechanism of friction, as the vibrating air transfers its energy to the walls and to objects within the room, trying to make them move in sympathy with the sound vibrations. Even the air itself absorbs some sound energy through frictional losses. However, it is important to understand that the actual amount of heat generated from the conversion of sound energy sound is always very small, and no one has yet managed to burn their house down just by turning their stereo up too loud!

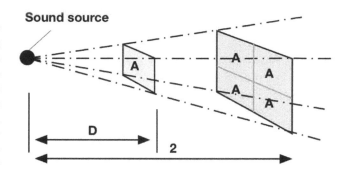

Sound source

The Inverse Square Law – as the sound wave radiates away from the sound source, the larger the spherical wavefront becomes, so the more thinly the sound energy is spread over its surface. D, the sound covers an area represented by the square of area A. After travelling twice the distance the same sound energy now covers an area four times that of A, so it now has only one quarter of the intensity for each A-sized square. Doubling the distance reduces the sound intensity by a factor of four, so if we were to double the distance to 4D, the energy reaching each A-sized square would only be one sixteenth of that at distance D.

The intensity of a sound wave reduces in level the further it travels away from the source, as its energy is spread over an ever-increasing surface area. Where the sound source is able to transmit equally in all directions, producing an expanding spherical wavefront, its intensity follows what we know as the 'inverse square law', which states that sound intensity decreases at a rate which is proportional to the square of the distance from the source. So as the sound wave radiates away from the sound source, the larger the spherical wavefront becomes and so the more thinly the sound energy is spread over its surface and the less energy there is at any one point on the surface to hear.

Sound Meets Wall

Sound is essentially rapidly repeating changes in air pressure occurring at frequencies that fall within the range of human hearing, which we put at broadly 20Hz to 20kHz. The vibrations are carried through our atmosphere, travelling at roughly 343 metres per second, at room temperature (the speed of sound changes slightly with variations in temperature and humidity). Sound can also be conducted via solids or liquids, although the speed of transmission will be different from that in air – and usually a great deal faster. The reason sound leaks from an enclosed room (let's assume no doors or windows for the moment) is that its vibrational energy causes the walls to vibrate in sympathy so that they in turn launch new sound waves on the other side of the wall and hence some of the sound inside the space can be heard outside too. Lower frequency sounds tend to carry more energy and cause greater sympathetic vibrations in the structure, and so it is the bass that tends to leak the most.

The more massive the wall, the less it will move in response to the sound, resulting in less sound being produced on the other side. For example, a dual layer 'cavity' brick wall, as used in much modern building construction, may attenuate the sound transmission by more than 50dB which is often all we need. Those of you thinking ahead might then make the intuitive leap that if one cavity wall can attenuate the sound transmission by 50dB, then building a second cavity wall spaced a little way from it might drop that sound by another 50dB giving us a fabulous 100dB of isolation in total... If only the laws of physics were so accommodating!

Unless the walls are separated by a large distance, the air filling the space between the two walls couples energy from one wall to the other, reducing the actual isolation to well below our hoped-for figure of 100dB. However, an air gap is still beneficial in reducing sound leakage by a useful amount, as double glazing aptly demonstrates. The wider the air gap, the better the isolation, especially at low frequencies, and a correctly built double-layer structure invariably provides better isolation than a single wall of similar mass. This is true even if the air space is just a few centimetres wide, although in commercial studios, corridors and machine rooms are often planned to form part of the sound isolating structure. The width of the corridor provides the air gap between the two walls, making the low-frequency isolation much more impressive.

However, in general, to reduce the amount of sound leaking out of a room we need to reduce the amount that the outside of the wall can move, and adding mass to the walls by using heavier materials or by making them thicker will reduce the amount of sound transmission. If we double the mass of a brick wall by doubling its thickness, the same amount of sound energy will only be able to move it half as far, so the amount of sound leakage will be halved, or reduced by 6dB. At this point you might be thinking that 6dB doesn't seem that great for all that extra mass, especially if you were hoping to reduce the sound leakage by something closer to 50dB! If you can build one-metre-thick heavy stone walls you might be in with a chance of that kind of reduction, but that's hardly practical in most cases.

A more practical alternative is to choose lighter materials that are more 'lossy' to sound transmission. For example, a layer of sand could be expected to transmit less sound than the equivalent mass of rigid material as the movement of the individual sand grains would cause more 'sound to heat' conversion via friction. Various commercial materials are available that include lossy layers, with the most effective solutions often combining high mass with a lossy structure. For example, using a sandwich of limp-membrane barrier mat (mineral loaded vinyl) or lead sheet in between drywall boards can be a very effective solution.

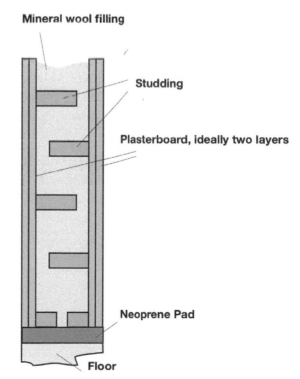

Mineral wool filling

Studding

Plasterboard, ideally two layers

Neoprene Pad

Floor

A simple plasterboard-on-studding construction, with mineral wool or glass fibre in the cavity to reduce resonances, can be highly effective. Where possible, the two frames should not be in contact with each other.

So if you need to create a new partition wall, but you don't want to use brick or concrete block, a plasterboard-on-studding construction is often adequate, ideally with both sides built onto separate frames that are not in contact with each other. Mineral wool or glass fibre can be used to fill the cavity to reduce resonances, while multiple layers of plasterboard on at least one side helps maximise the mass of the partition – but do check that the floor is strong enough first!

The bigger the air gap between walls the better the performance, and isolating the partition from the floor and adjoining walls using neoprene blocks will also help considerably, although all air gaps around the edges must be filled with a resilient material such as silicone rubber or

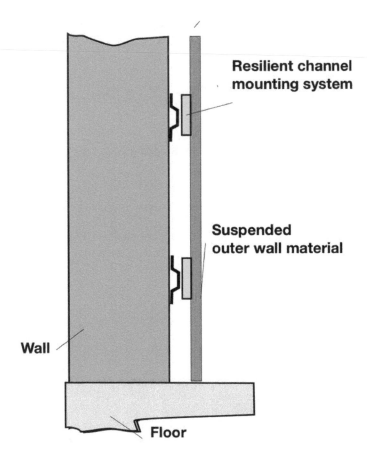

Resilient channel mounting system

Suspended outer wall material

Wall

Floor

◄

Simplified view of a resilient-channel suspension system. The channel flexes to absorb the sound energy rather than transmitting it through to the wall behind.

mastic. An alternative is to build a single-stud partition with plasterboard on both sides, and then create a further inner layer, using a decoupled mounting system. There are also commercial systems available that mount a lightweight inner wall on resilient metal fittings attached to the existing wall in a way that reduces the amount of energy transmitted through the whole structure. Many of these kinds of solution can also be fitted by any reasonably DIY-capable person. A flexible metal channel or series of isolation clips are fitted to the original wall, then the new drywall sheet is screwed to the channel or clips.

And The Bad News...

The degree of sound isolation provided by a structure reduces with frequency, so the most common sound leakage complaints usually relate to low-frequency sounds, such as kick drum and bass guitar. This is largely attributable to the mechanical inertia of the wall.

Since sound attenuation is frequency-dependent the attenuation of a particular material is generally measured in decibels at a number of specific frequencies between around 100Hz and 3kHz. This figure is called the Sound Reduction Index (SRI) and while most materials also come with an average specified SRI value, what you really need to know is how much attenuation you can expect at the lowest frequencies you're trying to isolate. Online data is available from many building material manufactures giving SRI figures at all the specified frequencies, so it is worth taking some time to browse these to see what you can realistically expect to achieve by using them.

The practical SRI of a material or wall structure is affected by a number of factors. We've already mentioned the importance of sheer mass and also the way the stiffness of the material affects how lossy it is. However, there's also the subject of resonance. If a partition exhibits a strong resonance whenever a specific note is played the sound energy will pass through the wall with far less reduction in intensity than other frequencies. Multiples of the resonant frequency will also be attenuated less. Lossy materials or multiple sandwiched layers of materials with different acoustic properties can help damp or remove these undesirable resonances.

In a domestic room with drywall/stud internal walls, the best compromise is usually to fit one or more additional layers of drywall. This works best with a further damping layer between the original and new sheets of drywall. Suitable drywall products are available with a resilient layer already bonded to the back, or self-adhesive damping sheets or specialist pastes are also available for the purpose. The idea of this constructional technique is that not only does the additional drywall layer add mass, but it also damps resonances and changes the frequency of such lesser resonances that do remain so that they don't coincide with those of the single drywall on the other side of the wall.

If you need to treat a wall which adjoins neighbours and the budget allows, then consider consulting a professional company with a view to fitting an internal wall suspended on a resilient channel system as these can be extremely effective in reducing the transmission of unwanted low frequencies, without having to add excessive amounts of extra mass to the wall structure.

The Weakest Link

While a solid wall can give around 50dB of isolation, a studding/drywall partition wall will provide significantly less, and a lightweight domestic door might struggle to give you more than 15dB of isolation. Likewise, windows, even double-glazed ones, provide far less isolation than a solid wall. You can make your own ball-park measurements by playing pink noise or a sound loop in your studio and using a sound level meter app on a smart phone to compare the sound pressure levels (SPLs) inside and outside the room.

The weakest areas for sound leaking in or out of most project studios are the doors and windows. Treating the walls to improve their isolation may be beneficial where one or more of the walls adjoins neighbours or rooms used by other family members, but usually the most immediate benefits will be gained by sorting out problems involving doors and windows. How effective the isolation needs to be depends both on your surroundings/neighbours and on the time of day (or night) you plan to work. Low-level noise leakage is often masked by ordinary daytime sounds such as traffic, but may be much more noticeable at night when the ambient sound level outside falls significantly. Often the answer to problem sound leakage out of a studio is to combine a practical level of soundproofing treatment with a pragmatic reduction in the sound level being made in the studio.

Whilst double-glazed windows offer only a limited amount of sound isolation when compared to a solid wall, they are far more efficient than single-glazed units, partially because of the dual-layer, air-gap structure, but also because they tend to have much better, airtight seals than single-glazed windows. Which brings me to a very important point: rooms that are sound-tight are also air-tight: just a 2mm gap under a door can leak an enormous amount of sound and so must be dealt with or it will negate the effects of anything else you have done to achieve isolation. You may have to open the door from time to time to allow in some fresh air, but that's a small price to pay for being able to actually work in your studio!

Window Upgrades

The performance of double-glazed windows can be boosted even further where necessary by adding an extra heavy-glass or acrylic-sheet internal window (or even a secondary double-glazing system) with as large an air gap as possible between it and the original double-glazed window. Heavy curtains can help a little and will also provide some absorption to help with the room's own acoustics, but they won't be very effective at low frequencies. Small windows, or those comprising multiple separate panels, work best because large sheets of glass tend to be quite resonant, although if you can fit a separate internal window using a different thickness of glass from the original double-glazed unit, the severity of any resonances can be reduced.

Where you don't need the daylight you can fill the window space with sandbags or high-density mineral wool slab and then board over it with a couple of layers of thick plasterboard or chipboard. Even in a rented room this is a practical

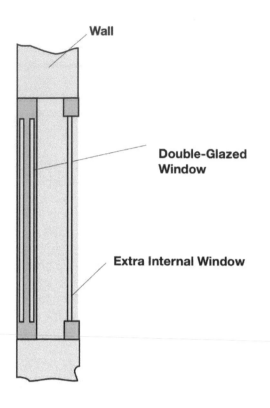

Wall

Double-Glazed Window

Extra Internal Window

Adding an extra heavy-glass or acrylic-sheet internal window, with as large an air gap as possible, will boost the isolation performance of any window, even a double-glazed one.

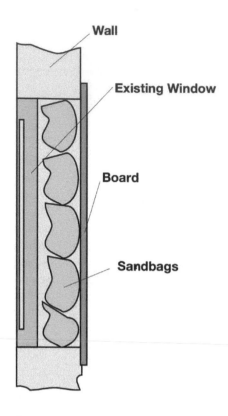

Wall

Existing Window

Board

Sandbags

Windows should be airtight and ideally either double-glazed or blocked off. Blocked off windows can be filled with sandbags to give much improved isolation.

temporary solution that will leave only a few screw holes to be filled when you move. Note that if the window you are boarding over isn't airtight, but you don't want to replace it, you can use some acrylic or silicone frame sealant applied from a mastic gun to seal around the edges first. In a rented property you can use gaffa tape to seal around the window frame.

Doors

While you can seal up unwanted windows, you'll always need at least one working door! Domestic doors are often made from two very thin plywood or MDF panels separated by a lightweight

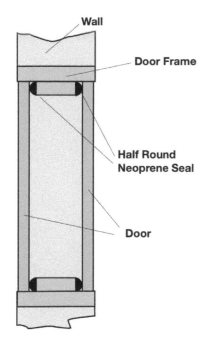

Wall

Door Frame

Half Round Neoprene Seal

Door

▲
Doors need to be heavy with airtight seals on all four sides. A heavy compression latch will keep the door closed tight. Double doors work even better, if you have the space to fit them.

TIP: Rather than fitting the wooden 'closing strips' to the door frame first, it is far better to glue your neo-prene seals to the wooden strips using contact adhesive, before pinning them in place. You can then precisely position the strips against the closed door so that the neoprene seal is just tight enough against the face of the door to hold a sheet of paper – any tighter and you may have problems closing the door. Ensure the corners are mitred neatly so there are no gaps in the seal and use a silicon sealer, if nec-essary, to ensure the corners are airtight.

frame, so they don't have the necessary mass to act as effective sound isolators – and they can also be very resonant. They are not designed with airtight seals either as this would not usually be a domestic requirement, but when it comes to sound isolation the door must be airtight on all four sides.

To add mass to the door the simplest and cheapest solution is to fix a layer of 12 to 18mm chipboard, MDF or even drywall board to both sides, although replacing a lightweight door with a commercial 'fire door' is also worthwhile as these are necessarily heavy. Some form of sealing strip can then be applied around all four edges to form an airtight seal when the door is fully closed (see below).

Where space permits an even better solution, and one that is employed in many commercial studio designs, is to use a double-door entrance lobby with an air gap between the two doors. In a commercial studio this may take the form of a large 'air lock', often combined with storage space for unused gear, but in a home studio even a space of only the wall thickness between the two doors will make a big difference. It is possible to modify most door frames quite easily to take two doors, one flush with the inside of the wall and one flush with the outside.

When it comes to installing airtight seals around doors, consumer draught-proofing isn't very effective for acoustic purposes, and we recommend using the kind of semi-circular, neoprene, sealing strip available from a studio-materials supply company instead. It is also best to fit the door (just one of the doors in a double-door system) with a compression latch that presses the door face firmly against the seals when closed. A typical domestic door latch won't usually keep the door pulled tight against the seal.

Commercial studio doors often incorporate a sealing strip on their lower edge that is automatically raised and lowered as the door is opened and closed, as this avoids having to have a raised threshold strip on the floor. However, the only practical solution when using modified standard doors or fire doors in a domestic situation is to employ a raised threshold strip to give the lower edge of the door something to seal against.

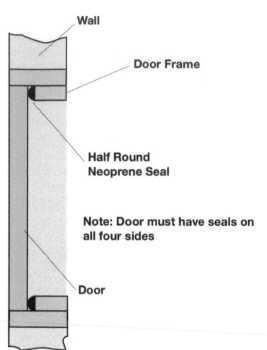

Wall

Door Frame

**Half Round
Neoprene Seal**

**Note: Door must have seals on
all four sides**

Door

◄

Typical consumer draught-proofing isn't very effective for acoustic purposes –
we recommend using semi-circular, neoprene sealing strip, available from all
good studio-materials supply companies.

Floors
. . . .

Suspended concrete floors can offer a reasonable amount of
sound isolation from airborne sound because of their significant
mass, but sound energy coupled into the floor through direct
physical contact will be transmitted with very little attenuation
at all. Consequently, it is usually necessary to provide some
additional means of isolating any direct source of vibration, such
as footfall, kick drum pedals, guitar amps and so on. Suspended
wooden floors generally offer poor isolation because of their lighter
construction and also because most are not airtight. The airtight
issue can be solved easily enough using flexible mastic filler, but to
provide a good level of contact-sound isolation is for wooden or
concrete floors to create a so-called 'floating floor' on top.

Professional floating floors can be extremely complex, some
comprising thick concrete structures suspended on springs,
but fortunately you don't need to worry about that kind of
expense for home use! A simple DIY technique that we've
tried – both for complete floors and for sound-isolating drum

plinths – is build a simple structure comprising two layers of 16mm chipboard glued and screwed together, resting on a bed of 30mm or 60mm thick mineral wool cavity wall insulation laid upon the original floor surface. The mineral wool should have a density of around 60kg per cubic metre or higher in order to provide sufficient stiffness to take the weight of the new flooring. Using two layers of board not only adds mass but also ensures the structure is rigid if one set of boards is arranged lengthways and the layer above orientated widthways (i.e. at 90 degrees to the first layer). If you aren't planning on carpeting the floating floor, then plywood may make a more durable and attractive upper surface. Alternatively you could fit standard laminate flooring on top of the chipboard.

Of course, this arrangement will result in a small step up into the room, and it will be necessary to modify the doors (and possibly the skirting boards) to fit above the new floor level.

It is essential to avoid vibrations from the floor coupling into the surrounding structure, so if installing a floating floor across the entire room, strips of felt or neoprene should be fitted around the walls to prevent the new floor from touching them. If the new floor was allowed to touch the existing walls sound would be transmitted via direct mechanical contact, essentially 'shorting out' the new resilient floor completely! Similarly, if you fit a skirting board you must leave a small gap below it so it doesn't touch the new floating floor layer.

If the room has a wooden floor and sound transmission to the room below is a problem, you may get a worthwhile improvement by laying 20kg per square metre barrier mat,

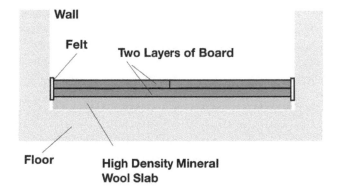

A basic DIY floating floor can be built with two layers of 16mm chipboard, glued and screwed together, resting on a bed of high-density, 30mm- or 60mm-thick mineral-wool cavity wall insulation.

sometimes called dead-sheet, on the floor before putting down the mineral wool. Barrier mat is a flexible vinyl material loaded with clay particles, and in this application it adds mass, seals gaps between the floorboards, and absorbs a useful amount of energy due to its 'lossy' structure. Most studio-materials suppliers have various types in their catalogues.

However, while the improvement in sound attenuation should be significant if you install a floating floor, it's still unlikely that you will be able to use an acoustic drum kit in a wooden-floored room without some of the sound still being audible in the room below. And even if you use an electronic kit, the thump from the mechanical pedal action still tends to come through, so a separate floating drum plinth, using the construction just described, is still a worthwhile addition.

Ceilings

Ceilings are more difficult to treat, as to bring about any serious improvement you need to construct a suspended ceiling below the original ceiling leaving as big an air gap as possible – and even then the results may not be as good as might be hoped for. Few DIY enthusiasts will want to tackle this job, and this is one area where you might want to think about calling in the professionals. However, if the room above is part of your own building, installing a layer of barrier mat (or other specialist noise absorbing underlay material) on the floor above, between the floor surface and the normal floor covering, will certainly help. In more extreme cases, building a floating floor in the room above might do the trick, so long as you don't mind the slight step up into the room and having to modify the doors and skirting to fit above the new floor level. Where the budget allows, there are also commercial systems for improving ceilings, some of which rely on sound-deadening panels fixed to flexible metal channel, similar to that used for walls.

Good Vibrations

Although we've already hinted at it, one area that we haven't discussed in detail so far is the potential problem of sound being conducted along solid materials in a building, sometimes known as 'flanking transmission'. Such vibration-borne sound can bypass or short-circuit all your isolation work. One obvious scenario would be where a loudspeaker is rigidly mounted onto a steel joist that passes through to the next room. Vibration, and therefore sound, will readily travel along the joist, circumventing any barrier in between the rooms – I'm sure you've seen those prison-escape films where the inmates pass messages by tapping on the water pipes! It's the same principle.

This is the reason why we have described isolating secondary walls and floating floors from the original room structure as much as is possible by the use of resilient mountings, neoprene pads, flexible hangers, flexible adhesive and flexible joint-sealing compound. Similarly, any noise-making gear in the studio, such as guitar amps, monitor speakers, drum kits and so on, should be isolated from the floor as best you can. The spiked feet on speaker stands can be important in ensuring that the stand is stable, especially on carpet, but those metal spikes will also conduct vibrations directly into the floor, necessitating the use of speaker-isolation pads or platforms to decouple the speaker from its support. Larger isolation platforms are also available for guitar amps, though you could make your own by gluing spare acoustic foam to the underside of a wooden board. Drums, including electronic kits, can be isolated from the floor by making a kit-sized section of floating floor, as we have already seen.

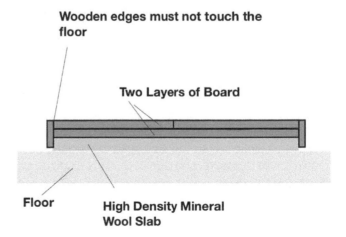

Wooden edges must not touch the floor

Two Layers of Board

Floor

High Density Mineral Wool Slab

> Noise-producing items should, where possible, be isolated from the floor or other parts of the structure. Commercial isolation platforms are available, but it is perfectly feasible to build your own.

Summary

Whilst you may not be able to get as much sound isolation as you'd ideally like, you will usually see a significant improvement if you address the weakest areas first, specifically doors, windows and any 'noisy' equipment placed directly on the floor. Start with air-tight seals around doors and windows, then re-evaluate the situation to see where the remaining weak spots are, tackling them in order of seriousness. After that, look to upgrade any particularly lightweight walls and consider building a floating floor. When you have reached the limit of what can be done practically or cost-effectively, you will need to compromise how you use the room, such as generally adopting with lower sound levels or only monitoring via headphones when working late at night.

Since a 'soundproof' room needs to be an airtight room, you also need to think about how to get fresh air into your studio. Full air-conditioning systems can be noisy, while ducted 'quiet' air systems tend to be costly and may also require space that you simply don't have. However, computers and outboard equipment nevertheless generate a lot of heat, which, in a well-insulated studio, may build up very quickly. The simplest and cheapest solution is to take regular breaks, leaving the door open to let the air circulate for a while! But perhaps a more practical and reasonably affordable compromise, (and the one that Paul uses in his own studio), is a basic air conditioner called a 'split system'. This involves an air cooler on the wall inside the room, with a linked heat exchanger outside the building. Although not silent, this kind of system only needs to be run for few minutes at a time to bring the temperature down. Whereas most portable air conditioners either need a large pipe running to the outside world (which will negate much of your sound isolation) to prevent them releasing a lot of potentially corrosive water vapour into your studio, split systems need only a small hole in the wall, which can be sealed with mastic once the installation is complete. Of course, cooled air isn't the same thing as fresh air, so you'll still need to open the door on a regular basis to ensure a change of air in the studio to bring in some new oxygen!

Chapter Six
Cables and Connections

There's nothing quite so dull as the subject of studio wiring when all you want to do is make music, but if you don't pay sufficient attention to this important topic, your system may not perform as well as it should, as we've confirmed on a number of our *Studio SOS* visits. Even plugging into the wrong mains sockets can cause unexpected problems, as can having too many or too few ground connections. To illustrate some of the typical problems and their solutions, we'll start by examining some real-life examples we've encountered.

The importance of appropriate mains connections was very aptly demonstrated when we received a *Studio SOS* distress call from a reader whose audio system muted for a second or two whenever anyone switched on an appliance or even a light elsewhere in the house! We eventually traced the problem back to a digital mixer he was using for monitoring, fed from the digital output of his interface, and the mixer was upset whenever other electrical items were powered on, or the central heating boiler fired up. His studio system was powered from two wall sockets on opposite sides of the room. We weren't sure whether these were connected to the same ring-main or whether they were on separate circuits, but problems can arise very easily when systems are powered from two or more different power points because the socket grounds take different paths to the main building ground point, potentially creating 'ground loops' which can induce mains-frequency hum into the system. (See p. 95)

The simple solution in this instance was to redo the studio powering arrangements so that the entire studio ran from just one wall socket. The amount of current taken by a typical project studio is far lower than the maximum capacity of one power socket, so there are no safety problems associated with doing this.

Our strategy was to fan out all the power points from this one socket using a surge-protected, six-way distribution board, which in turn fed a number of conventional four-way distribution boards to provide the necessary number of outlets for all the equipment. In effect, the mains power is distributed in a star arrangement, and the equipment mains safety grounds are all returned directly to the same wall socket via the shortest possible path. It makes sense to plug the most power-hungry devices into sockets as close to the wall outlet as possible.

Everything was re-connected to this new power system, including the keyboard and electronic drum kit, and when we retested it the hum-induced digital-dropout problem had vanished completely. No changes to the signal cabling were

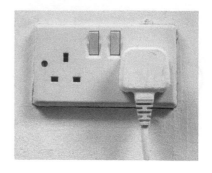

The amount of current drawn by a typical project studio is far lower than the maximum capacity of one domestic power socket, so there are no safety problems associated with running everything from a single outlet.

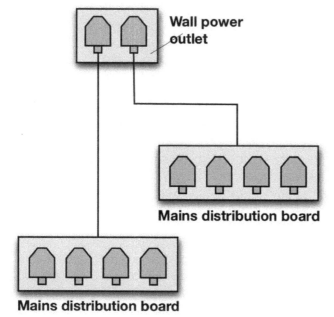

Wall power outlet

Mains distribution board

Mains distribution board

All equipment power 'fanned' out from one wall socket (or pair).

TIP: In analogue systems, ground loops are usually audible as unwanted background hums, buzzes or whining, but in digital systems the effects don't always produce an audible artefact at all. Instead, they may simply compromise the way data is transferred from one device to another, resulting in occasional, unexplained clicks or drop-outs. Note that optical digital connections can't create ground loops, although a loop may still be present in the system via other, wired connections.

So long as you are not exceeding the current rating in any part of the chain, there's nothing wrong with cascading power distribution boards from the same outlet in a star arrangement.

required, as there was very little signal wiring and the cables were very short, other than the feeds from the keyboard and an electronic drum kit.

/

Ground loops can arise in any moderately complex audio system where there are multiple ground or earth paths between pieces of equipment and between the equipment and the mains power-supply safety earths. Take the example of a mixer connected via an unbalanced cable to a powered monitor, with both devices plugged into different mains wall sockets, several feet apart. The mains safety-earth connections from those two sockets are joined together at the building's fuse box after many meters of cabling winding its way through the building. The 'ground loop' is completed by the signal-cable ground connecting the mixer and the monitor.

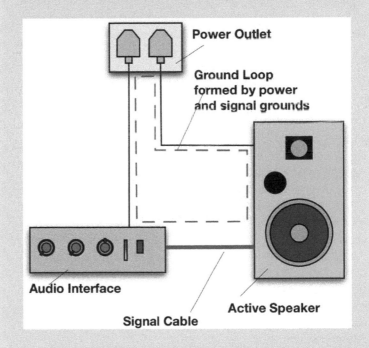

Power Outlet

Ground Loop formed by power and signal grounds

Audio Interface

Signal Cable

Active Speaker

Ground loops can occur whenever there are multiple ground paths created by the signal cabling between items of equipment and their mains power-supply safety earths, causing audible hum and buzz, or digital instability.

Most equipment incorporates some form of mains filtering which dumps unwanted noise into the safety earth in the form of small 'leakage' currents. The small but finite resistance of the mains earth cabling means that the small leakage currents flowing along it generate a small and varying voltage across the entire earth cable, so that the actual voltage at the mains wall socket earth terminal might be slightly above the nominal 0 Volts. This has no bearing on the safety role of the socket's earth terminal, but ground loops can become problematic when that voltage is allowed to couple back into the equipment's audio circuitry, since it then becomes audible as hums or buzzes, or causes digital instability.

The surefire way of providing that coupling is to allow a current to flow in a closed circuit through the equipment's audio grounds. In the example above, the loop exists between the earth of the first mains socket, through the plugged-in equipment, along the grounded unbalanced audio cable, through the connected equipment, down to another earthed mains socket, and then via the building's wiring to the central earth termination where it then finds its way back up to the first mains socket to complete the loop!

If a system is fed from two or more power sockets that happen to be on different ring main circuits, as we suspect may have been the case in the preceding *Studio SOS* example, the size of the ground loop (or loops) may be many tens of metres, running all the way back to the fuse box or consumer unit. As a rule, the more cable in the ground loop, the larger the induced ground voltages and the better it is at picking up interference. Adopting a fan or star-shaped mains distribution system fed from a single power outlet helps to minimise ground-loop problems by keeping the ground paths as short as possible while also ensuring a common ground reference point.

Disconnecting one or more mains cable safety grounds may break a ground loop and appear to solve a hum problem, but it is most definitely not a good idea from the safety point of view. A serious equipment fault could well become lethal in

TIP: Don't run signal cables alongside power cables, if at all possible, as this can also result in magnetically induced hum. If audio and power cables have to cross, making the crossings at right angles will minimise the possibility of interference.

Keeping your mains and signal cable runs separate will not only assist when trying to diagnose a fault, but also reduce the possibility of accidentally inducing noise.

such circumstances. Ground loops must only ever be resolved by interrupting the audio cable screen in some way, as will be explained shortly.

double-insulated (Class II) power supplies (wall warts and line lumps!), which incorporate isolating transformers, without a mains-ground connection any-where. This meant that the whole system was 'floating' and the screens of all the connecting cables were effectively acting as radio aerials, picking up all sorts of unwanted rubbish from the ether and passing it all directly into the audio electronics.

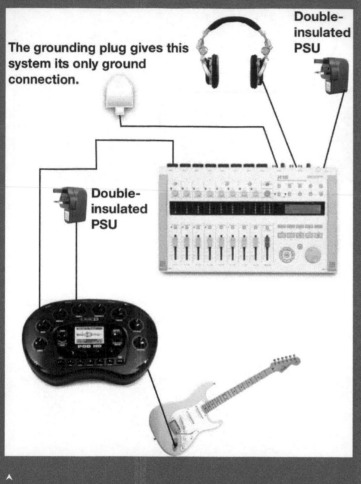

The grounding plug gives this system its only ground connection.

Double-insulated PSU

Double-insulated PSU

If everything in your recording system uses wall-wart or line-lump double-insulated (Class II) power supplies, which incorporate isolating transformers, you can end up without a mains-ground connection at all!, resulting in excessive hum. The problem can be solved using a dedicated grounding mains plug, such as the one opposite (note the model shown is for a UK 13A socket).

The solution was to attach a length of wire to the metalwork of the Zoom recorder via one of the screws on the underside, and to ground the other end to the metal chassis of a piece of household equipment that was properly grounded – in this case, a nearby hi-fi system. As soon as we did that the hum problem vanished entirely because the cable screens were then able to act as proper screens, trapping unwanted interference and passing it to ground rather than into the audio circuitry!

In theory you could create a ground cable by wiring only a ground wire to the earth pin of a standard three-pin plug, but we certainly don't recommend this as tripping over the cable might just disconnect it from the ground pin in the plug and allow it to touch the live pin, which would then cause all the connected equipment to become electrically live. A far safer solution is to buy a dedicated grounding mains plug fitted with a metal ground pin and plastic dummy pins for the live and neutral. Although these are not stocked by typical DIY stores, an on-line search for "grounding mains plug" brought up several sources.

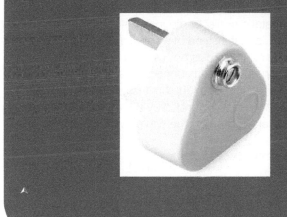

Balanced and Unbalanced

Audio-signal connections, including digital ones, can either be balanced or unbalanced. An unbalanced audio connection employs on a two-conductor cable, with an insulated inner core surrounded by a conductive screen, that also forms the return path for the audio. This is how a typical guitar cable is constructed. Although the screen (which is generally connected to ground within the connected devices) offers a degree of protection against electrostatic interference, it is still possible for traces of outside interference – especially electromagnetic interference – to become superimposed on the wanted audio signal, if the supposedly ground-potential screen starts to pass small AC currents (as can be the case when a ground-loop occurs). Although the current passed is far too small to be dangerous, it can result in the screen voltage being modulated sufficiently to cause audible hum.

In any system using unbalanced audio connections, it is inherently very difficult to keep the ground currents out of the audio circuitry, simply because the cable screen also functions as the signal-return path and any hum induced into the screen automatically becomes part of the signal.

Balanced systems were developed to allow audio signals to be carried over longer distances without being significantly affected by electrostatic or electromagnetic interference. The underlying concept is that instead of a two-conductor system that uses the same conductor as both screen and signal-return connection, as in an unbalanced connection, a balanced cable employs two 'floating' signal conductors (usually referred to as the 'hot' and 'cold' wires), surrounded by a separate screen that carries no signal – it is purely there to provide electrostatic

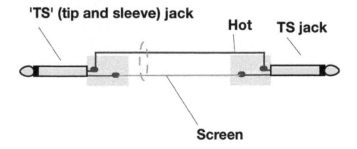

'TS' (tip and sleeve) jack

Hot   **TS jack**

Screen

In an unbalanced cable, the screen doubles as the electrical return or 'cold' signal path.

screening. There is no direct connection from either signal wire to ground: both have the same reasonably high resistance to ground and hence 'float' away from ground.

Systems using balanced connections, where the cable screen is not involved in conveying the audio signal at all, *should* be immune to ground loops... although not all designs are!

Whilst a well-screened cable can actually be very effective against electrostatic interference, screening alone isn't very effective against *electromagnetic* interference – a problem that the balanced system solves in a simple but ingenious way. The two audio signal wires are twisted tightly together and connected at the receiving end to a 'differential amplifier' – an amplifier with two summed inputs, one inverting and one non-inverting.

Any electromagnetic interference that makes it through the cable screen will induce identical voltages in the two signal wires – this is guaranteed because both wires have the same impedance to ground, and the tight twist means that both wires are exposed to the same magnetic field. When these identical 'noise' voltages reach the differential amplifier they essentially cancel each other out, leaving only the wanted audio-signal voltage.

Often, the 'cold' signal wire carries a polarity-inverted version of the audio signal on the 'hot' signal wire, with the difference between the two – which is what the differential amplifier is interested in – therefore being twice the audio signal on either wire individually. However, balanced interconnections don't *have* to carry symmetrical polarity-inverted signals on the hot and cold wires – that aspect plays no part in the interference-rejecting properties of the interface.

A lot of equipment these days uses 'impedance balanced' connections in which the 'cold' wire carries no signal at all. This technique is definitely still balanced because both signal wires have the same impedance to ground, which ensures the matched interference voltages that the differential amplifier relies upon for cancellation. The interference-rejection properties and freedom from ground loops associated with a symmetrical interconnect is maintained. However, this arrangement is both cheaper to implement, and has a very useful practical advantage when working with both balanced and unbalanced destinations because the full signal is present on the 'hot' wire. So an unbalanced destination that only connects to the hot wire and cable screen will still see the full signal. In a situation where the balanced interface splits the signal between the two wires using the symmetrical polarity-inverted approach, an unbalanced destination would potentially only see half the full signal, and so would receive a signal level 6dB quieter than it should be!

XLR connectors are often used in professional circles for carrying balanced signals, while TRS (3-conductor, 'tip, ring and sleeve') jacks are also used in project-studio gear. The international standard XLR wiring protocol is to use pin one for the screen, pin two for the 'hot' signal and pin three for the 'cold' – easily remembered as **X**reen-**L**ive-**R**eturn ('XLR' equals 1, 2, 3). Balanced TRS jacks are wired 'tip hot', 'ring cold' and barrel, or sleeve, to screen.

In 'class 1' balanced audio equipment, the metalwork of the case is grounded to the mains safety earth and the pin 1 screen of the XLR connectors should also be connected directly to the case. In this way the audio electronics are completely and properly shielded from external interference, and any ground currents in the cable shields are kept well away from the audio electronics. Sadly, some manufacturers don't adhere to the correct grounding procedures and the cable screen is often connected to the audio-amplifier ground-reference point, rather than directly to the case. This arrangement effectively injects unwanted ground-loop hum straight into the audio electronics. One way of avoiding ground-loop problems in these cases is to disconnect the signal screen at one end of the cable – usually the destination end. Since the screen connection plays no part in conveying the audio signal in balanced cables, this solution is perfectly acceptable in most cases, but it obviously won't work with unbalanced cables. For those we can offer other solutions. Disconnecting the screen at the destination end potentially degrades the protection against electrostatic interference, but the improvement in ground-loop hums usually far outweighs that!

The wanted audio signal is applied differentially across the balanced line

Differential receiver ignores common-mode interference but passes differential audio

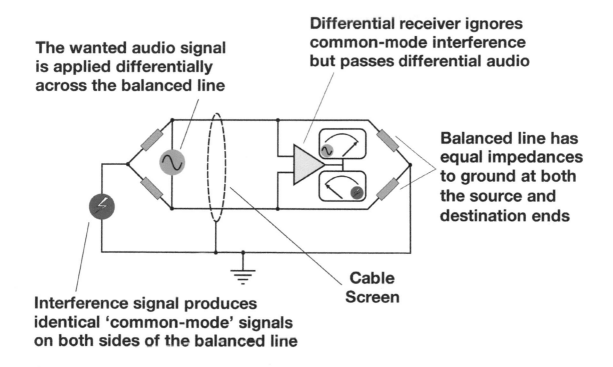

Balanced line has equal impedances to ground at both the source and destination ends

Cable Screen

Interference signal produces identical 'common-mode' signals on both sides of the balanced line

Any electromagnetic interference that makes it through the cable screen will induce identical voltages in the two signal wires. This is guaranteed because both wires have the same impedance to ground, and that's specifically what makes a 'balanced line' 'balanced'. When these identical interference voltages reach the differential amplifier (which subtracts the signal on one input from that on the other, and only passes the difference between them) at the destination they essentially cancel each other out, leaving only the wanted audio-signal voltage which is transmitted differentially between the two signal wires.

Some digital connection systems, such as coaxial S/PDIF, are not balanced. Optical S/PDIF cannot be responsible for creating ground loops as there is no signal ground in an optical cable. AES3-format digital interconnects are symmetrically balanced and should be immune to ground-loop problems.

Cable Screens

Not all screened cables are created equal – something that can become very evident when connections must be made using unbalanced cables. Even if ground loops are not responsible for inducing hum, a poorly screened cable may still allow hums and buzzes to creep into your audio. Cheap, moulded-plug cables, especially the very thin ones, often have a very loosely wrapped, spiral wire screen that still allows through a lot of interference, whereas a proper woven-screen cable is much more effective. They may seem to work OK in an area where the electrical supply is clean and there are no other sources of

▲
Wound and braided screens.

▲
Braid and foil screen.

▲
Foil screen with drain wire.

▲
Conductive-plastic screen with drain wire.

▲
There are different types of cable screen to suit different applications, with each offering different screening and handling attributes.

potential interference, but when the going gets tough, they're simply not up to the job.

In addition to woven-screen cables, foil screened cables are also excellent at rejecting interference, although they are more often used for permanent-installation wiring. Conductive-plastic screened cables don't work quite as well as a good woven-screen cable, but they are still perfectly adequate in most situations and have the advantage that they are very flexible, they don't kink, and they are easy to coil, making them ideal as instrument interconnects. There is little evidence that buying really costly, esoteric cables provides any further improvement in screening – indeed, some seem to have worse screening than standard cable types – but it pays to avoid cheap cables, as both the cable itself and the connectors on either end may be less than optimal.

Unbalanced to Balanced

The simplest way of interfacing unbalanced and balanced systems, while simultaneously avoiding ground loop problems, is to use a line transformer box. The transformer passes the signal magnetically between its separate input and output windings – which avoids any direct electrical connection and thus no ground loops – and it translates automatically between unbalanced and balanced connections too. It is important to use a unit with a transformer designed to handle line-level signals, rather than microphone-level signals, for this purpose.

However, if you have an unbalanced signal source, such as a synthesizer sound module, that you wish to connect to a balanced input, you can also avoid ground loops by taking advantage of the balanced input's differential nature. The *Sound On Sound* shop sells unique and bespoke-designed high quality 'pseudo-balanced' interconnects, but if you are handy with a soldering iron you can make a slightly simpler version. This special cable uses standard balanced (two core plus screen) cable, with a TRS jack or XLR connector at one end wired for normal balanced operation (hot connected to pin 2/tip, cold connected to pin 3/ring, and the screen connected to pin 1 or the barrel/sleeve of the TRS plug).

Note that at the unbalanced end the screen is left unconnected while the 'cold' conductor is attached to the plug's ground terminal

DIY 'unbalanced source to balanced destination' cable – the screen connects only at the destination end (or via the optional components).

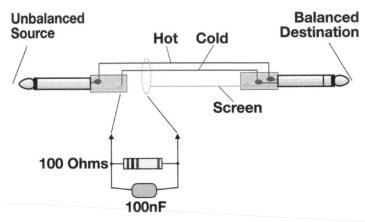

Unbalanced Source

Hot **Cold**

Balanced Destination

Screen

100 Ohms

100nF

Optional resistor and capacitor between jack ground pin and cable screen

At the unbalanced end don't connect the screen to anything – cut the screen short and tape over it, if required, to ensure it stays insulated. Connect the hot wire to the tip of the unbalanced TS plug (2-pole, 'tip and sleeve') and the cold wire to the barrel or sleeve. The 2-pole, unbalanced plug goes into the signal source output and the balanced 3-pole TRS-jack or XLR end plugs into the balanced input you're feeding, such as a mixer or audio interface. This arrangement transfers the audio signal perfectly, but avoids any direct ground connection and thereby avoids any possibility of a ground loop. Better interference rejection can sometimes be obtained by coupling the screen to the sleeve at the unbalanced end via a simple resistor/capacitor combination, but this is quite fiddly to achieve. The high-quality, ready-made pseudo-balanced cables available from the *Sound On Sound* online shop have these components built-in to maximize interference rejection, and extremely effective they are, too!

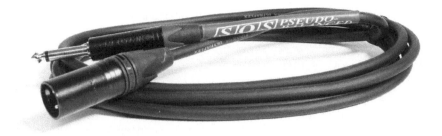

If you are not so handy with a soldering iron, you can buy high-quality, ready-made pseudo-balanced cables from the *Sound On Sound* website – www.soundonsound.com/shop

Gain Structure

Often we get calls for advice in situations where the user is having problems with excess noise or hiss. Most modern equipment is actually very quiet and, with the exception of guitar distortion effects, hiss shouldn't be a problem as long as you set all your levels correctly. This all comes under the broad heading of 'gain structure', and the short version of the mantra here is that all signals should be adjusted so they are at the nominal working level of the equipment into which they are feeding, well above the noise floor, but never allowed to get so high as to push the device into clipping. The latter, especially, is doubly important if the receiving device is digital. The nominal working level in most professional equipment is usually about 20dB below the clipping level, and this region is called the headroom margin. Semi-professional equipment might have a more limited headroom margin of maybe as little as 12dB. The system noise floor is typically around 95dB below the nominal working level in professional equipment, and perhaps a little less in semi-pro gear.

This approach to gain structure isn't applicable just at the input of your system – you should ensure that every piece of gear in your studio is running at its optimum signal level. If your signal level is set too low, you'll have to make up the gain elsewhere in the system, and gain always increases the level of noise as well as the level of wanted signal. At the same time, if the signal is too high you risk increased distortion and clipping. A lot of budget equipment can sound quite 'hard' and 'strained' when working with very high signal levels, and just optimizing the level around the nominal working level can often make a significant improvement in sound quality.

Good gain structure starts with the microphone preamplifier of your mixer, recorder or audio interface, and if you have a mixer equipped with PFL (Pre Fade Listen) buttons you can use these to check the levels of the individual inputs. Adjust the input gain trim controls until you get a sensible signal level on the meters, with the signal averaging around the 0VU mark (if analogue) or about –20dBFS on digital systems. The loudest transient peaks will typically be 8–10dB higher, although the sluggish response of an analogue VU meter probably won't show that at all!

In situations where a separate mic preamp is being used with an audio interface that also has its own line level gain control, you should first set up the mic preamp so that its own metering shows you are in the right area, then adjust the audio interface gain to get the right meter reading in your audio software, leaving 12 to 18dB of headroom if recording at 24-bit resolution. If you have a system with outboard equipment such as reverb units connected via analogue cables, ensure that you are also sending them sensible levels so you don't have to crank their input controls too near the top or the bottom of their range to get their meters reading correctly.

Life is somewhat easier if you are using an audio interface with its own mic preamps, connected to a computer running DAW software – here, you only need to adjust the mic amp gains until the DAW level meters show the correct level, again leaving a suitable amount of headroom as a safety margin. Distortion caused by clipping can't be undone without recourse to extremely costly restoration software, so is to be avoided at all costs.

Bear in mind that digital metering systems show the entire headroom margin all the way up to the clipping point (0dBFS). No analogue meter has ever done this – a VU meter typically stops at +3dB, and ignores the 17dB of headroom above it! Just because the digital meter shows that headroom margin, we are not obliged to use it – it is actually there as a safety margin to cope with unexpected loud transient peaks, the same as it was for analogue systems. It can sometimes look a bit strange to have all that 'empty space' above the digital meters, but there is nothing wrong with tracking and mixing with the average level around –20dBFS. In fact, that is exactly the right way to be working, and your analogue monitoring chain and outboard gear will thank you for it because they are all expecting to work with a 20dB headroom margin!

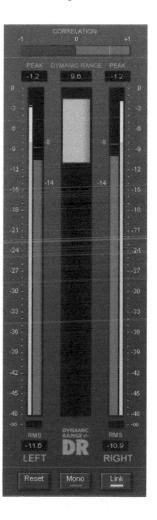

The metering in an analogue system generally shows only the normal operating level, with perhaps 20dB of safety headroom still available above this, whereas a digital system meter will show the entire range up to the point where clipping will occur.

Fully mastered material no longer requires a headroom safety margin because the peak levels are known, and so it has become standard practice to set peak levels of mastered material close to, or even at 0dBFS – but that doesn't negate the need for it when tracking and mixing.

Impedance

Impedance is essentially a circuit's 'resistance' to an alternating current (AC), but unlike resistance where a 100 ohm resistor still looks like 100 ohms whether the applied signal is 20Hz or 20kHz, impedance varies with the frequency of the applied AC signal. Of course, an audio signal is a perfect example of an AC signal that varies in frequency.

The higher the resistance in a circuit, the more voltage you need to create a given current flow through it, as dictated by Ohm's Law, which you may (or may not) remember learning at school. Using the simple Ohm's Law formula: $V = IxR$, you can always calculate the resistance, current or voltage in a circuit provided that you have the other two values (and don't mind a little algebraic rearrangement!). If you haven't encountered this before, R is obviously resistance (in ohms), V is the applied voltage and I is the current (in amps)

In a purely resistive circuit, resistance and impedance are the same, but in the real world circuits also have capacitance and inductance that impede the flow of current to a different extent at different frequencies. Don't panic! We're not going to introduce any more maths, but it is just worth knowing, at a conceptual level, that the impedance of a capacitor gets lower as the frequency increases, while the impedance of an inductor gets greater as the frequency increases. That's why these two particular components are so useful in constructing EQ circuits, and for dealing with unwanted external interference in audio amplifiers.

Most audio equipment is designed so as to keep the input and output impedances as constant as possible over the audio range. Loudspeakers, however, are a special case, as their impedance varies quite considerably with frequency.

The term 'input impedance' equates to the electrical load that the circuit presents to any source connecting with it – somewhat counter-intuitively, the lower the input impedance, the more difficult a load it presents because it takes more current to drive it. Output impedance is related to how much current an output can supply, and the lower the output impedance of a piece of gear, the more current it can supply – so the happier it will be feeding into a low-impedance input stage. Maximum power is transferred when the output impedance of the source is equal to the input impedance of the receiving device, and this is important in power amplifiers and loudspeakers. This 'constant impedance' or 'matched impedance' arrangement is also important for digital interfaces because at the very high frequencies involved cables behave slightly strangely, and constant impedance interfaces minimize internal signal reflections.

However, in the case of analogue microphone, instrument, and line connections we're not concerned with power and only care about the transferred signal voltage. Most of the time we simply want to get a signal out of one piece of equipment and into another without degrading it and the best way to do this is to ensure that the receiving device doesn't place too great a load on the sending device. Consequently, standard practice is for the source impedance (the output impedance of the equipment sending the signal) to be arranged to be between five and ten times lower than the input impedance of the device receiving the signal.

For example, a mixing console has a typical microphone input impedance of between 1,500 and 2,500 ohms, and the mic you might be connecting to it will usually have an output impedance in the region of 150 to 200 ohms. Line inputs have a typical impedance of 50,000 ohms, while line output stages often have an output impedance of 100 ohms or less. An additional benefit of this loosely adopted standard is that one output can often be split to feed several inputs at the same time, without any significant reduction in signal level.

On the Level

There are two 'standard' analogue operating levels in common usage: the –10dBV standard for semi-pro and

TIP: While you can split a signal to feed two or more destinations, provided that the combined load impedance isn't too low to be driven by the source, you can't just combine audio signals by simply joining outputs directly together. If you did this, each output stage will try to feed current into the others, resulting in distortion and possible equipment failure. We've come across instances where people have tried to do this as a means of mixing line level signals and in most cases, while nothing actually broke, the audio quality suffered significantly.

TIP: Most recording equipment operating at line level is designed to interconnect without matching problems, so in the normal course of events you won't need to worry about impedance matching. Just plug in the correct cable and you're in business.

◄

The two 'standard' analogue operating
levels: –10dBV for semi-pro and domestic
equipment, and +4dBu for pro gear. The
nominal semi-pro signal level is about a
quarter of the pro signal level, but so long
as you set your gain controls correctly, the
two standards can usually be made to work
together.

domestic recording equipment, and the +4dBu standard for
pro studio gear. The capital 'V' and lower case 'u' define
different signal level references (1 volt and 0.775V rms,
respectively, if you're interested), and if we convert these
levels to proper AC voltages, –10dBV is 0.316V while +4dBu
is 1.228V. In other words, the nominal semi-pro signal level
is about a quarter of the pro signal level. In decibel terms,
the +4dBu signal is roughly 12dB louder than the –10dBV
signal, but the reality is that these are close enough to each
other that, so long as you set your gain controls correctly, the
two standards can usually be made to work together without
much problem – most equipment has enough adjustable gain
range to handle either. In theory, connecting professional to
semi-pro equipment or vice versa will compromise your gain
structure slightly – in particular feeding a pro signal into a
semi-pro device will eat well into the headroom margin.
Going the other way will compromise the noise floor slightly,
but with well-designed equipment the deterioration in signal-
to-noise ratio is insignificant. However, where equipment
offers both –10dBV and +4dBu interfacing options you
should ideally use whichever is most suitable for feeding the
next device in the signal chain, even if it means obtaining
new cables.

Speaker Cables

Passive speakers should be connected to their power amplifiers using good-quality, low-resistance cables, with both speaker cables being the same length and as short as is practical. We've never found any conclusive differences in using expensive, exotic cables, but we have come across situations where the speaker cable used has been thin, lighting flex. This puts a significant resistance in series with the speaker, which will reduce the 'damping factor' of the system resulting in a less tight bass sound. Heavy-duty, twin-conductor mains cable, as used for garden lawnmowers, gives excellent results for the price.

For active speakers the signal is passed via ordinary line-level signal cables. Using balanced connections is beneficial where possible to minimize the risks of ground loop hums. Of course, active speakers also need mains power and there may be some benefit in buying heavy duty mains cables, as the ones

Using good-quality, low-resistance mains leads with high-current capacity is quite sufficient in any normal recording application, in our opinion, despite the great claims made for certain boutique cable designs.

supplied with equipment sometimes use thin cable, attached to the connectors at either end using crimps, rather than soldered or welded joints. A poor cable will form a resistive 'bottleneck' between your studio mains wiring and the equipment to which it is connected, so at the very least buy or make up a mains leads using high-current cable. The jury is still out over whether the more costly boutique mains cables can offer any significant benefit, although there's no doubt that they offer a low-resistance path, and may also have a filtering effect on high-frequency interference.

Cleaning

Connectors are made from metal and, with the exception of pure gold, they tend to corrode with age and with exposure to airborne pollutants – even gold can suffer from a build-up of contaminants. Corrosion and contamination can cause audible crackles, distortion and even brief dropouts. A specialised contact cleaner spray that does not leave a greasy residue should be applied directly to problem connectors, such as those in patchbays, XLRs or guitar jacks. This also works on mains plug pins where the ground pin, in particular, needs to be kept as clean as possible to ensure good conductivity. For cleaning jack plugs, the cleaner can be applied to a cloth for wiping the plugs or may again be sprayed directly on to the surface. In many cases, cleaning away the dirt and oxidisation results in a cleaner-sounding audio signal, and it certainly improves reliability.

The metal contacts in connectors can corrode with age and with exposure to pollutants – to date we have found no better troubleshooting contact-cleaner product than Caig Deoxit D5.

Summary

In most cases, a studio installation will perform at its best when all the necessary mains power connections are fed from the same ring main and from sockets that are physically close together. Using power distribution strips to fan out the mains supply from a single socket or a double socket is usually fine, as project studio equipment draws relatively little current. If possible, you should avoid using the same ring main that feeds your central-heating boiler, refrigerator, or any other heavy-grade electrical equipment, as these can result in audible mains contamination.

Where everything in your system uses 'double-insulated' mains connections (which have no ground connection), such as laptop power supplies, small portable recorders, guitar preamps and so on, you may need to create a dedicated system ground by connecting the screen of one of your cables (or the metalwork of the associated unit) to a properly grounded point using one of the methods described earlier in this chapter. If any one piece of equipment in your system is grounded, then interference problem should not arise.

If possible, use balanced audio interconnects and use either transformer line-isolating boxes or our unbalanced-to-balanced cabling solution when feeding unbalanced sources into balanced destinations. Use good quality screened cable, especially for longer cable runs, but keep all signal cables as short as is practical and don't run them alongside mains cables. Some basic soldering experience is always useful, as it allows you to make or modify cables to the correct lengths. Clean your connectors, including mains plug pins, every now and again to remove the grease and oxide that invariably builds up on them.

Chapter Seven
Vocal Recording

Getting a good vocal sound is easier than you might think, although it seems to be one of the subjects that people have difficulty with. You don't need fabulously expensive mics or preamps to get the job done to a high standard, so long as you follow a few basic rules. Large-diaphragm capacitor microphones, including suitable back-electret

Large-diaphragm capacitor mics are usually first choice
for vocals, but you can get perfectly acceptable results
with dynamic models too.

models, are most often used in this application because of
their ability to capture high frequencies more accurately, but
a good dynamic mic designed for live stage use will still work
perfectly well, and is even sometimes preferred for some
styles, such as heavy rock or rap.

Mic Powering

Capacitor mics always need a power source to operate,
both to polarise the capsule element and to power the
electronics of the on-board impedance converter that they
require. If you have a mic that can run on an internal battery,
it will usually be a type of capacitor mic called an electret,
which has a permanently pre-polarised capsule – the power
isn't needed to bias the capsule, but is still needed to run
the impedance converter. Tube (or valve, as we tend to
say in the UK) microphones need much higher voltages
to operate than other types of capacitor mic, and usually
have a dedicated power supply unit. Most studio mics,
however, can be 'invisibly' powered via the same cable that
carries their output signal – this is where the term 'phantom'
power comes from – so, if you want to use capacitor mics,
you need to have a preamp or mixer that can supply the
necessary voltage (48V DC is the standard, but lower-voltage

◄

Capacitor mics always need a power source to operate – either an external high-voltage supply in the case of tube mics, an internal battery for many electret mics, or a special 'phantom power' supplied invisibly via the balanced mic cable that also carries the audio.

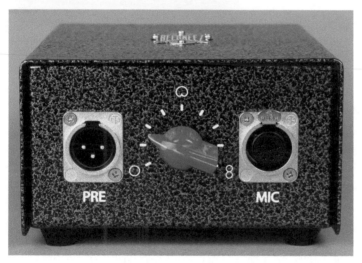

phantom systems are sometimes employed, mainly in older equipment).

'Boxy-sounding' vocals are often identified as a particular problem area, with studio owners frequently thinking that upgrading their mics or preamps will fix the problem. In the vast majority of cases, however, the real problem lies in the acoustic environment in which they are recording. Any unwanted 'roominess' in the sound at the recording stage will be further emphasised by any compression added at the mixing stage, so even what seems initially to be a relatively small amount of colouration can still become a significant problem when you come to mix. The simple fact is that even budget mics and preamps can give great results on vocal recording, provided you are working in a good acoustic environment. Fortunately, creating the right acoustic for recording vocals is both simple and far less expensive than upgrading your mics and preamps. Once you've got your room controlled in an appropriate way you might be in a better position to judge if more expensive gear will further improve your recordings. All you need to do at this stage, however, is to find a way to stop the room interfering with your recording. Fortunately, we have arrived at a cheap and virtually foolproof solution.

Directionality

Most people use cardioid pattern microphones when recording vocals, and these will, to some extent, reject unwanted sounds or room reflections from behind them. However, the cardioid pattern is only slightly less sensitive at the sides than the front (typically only about 6dB or so), so you can't rely on the pickup pattern alone to deal with all the unwanted room reflections. While it might ignore reflections coming directly from the rear, it will capture everything from the sides, above, below and from in front. This last point is particularly important because strong reflections from whatever is directly behind the singer, will bounce straight into the most sensitive 'business end' of the microphone.

So, the initial aim should always be to prevent as much as possible of the sound being reflected from the walls and ceiling from getting into the front and sides of the microphone, and one of the cheapest and most effective ways to do this is simply to hang a heavy polyester-filled, king-size duvet (or continental quilt) to form a concave U- or V-shaped curtain behind the singer. We don't recommend using feather duvets, as they cost rather more and, even more significantly in this application, all the feathers tend to sink to the bottom when they are hung upright! You can also hang duvets directly from the walls of the room to form a U- or V-shape across one corner, but try to keep as far out from the corner as you can to avoid bass build-up problems. You'll always get better absorption to a lower frequency if you can space the duvet away from the wall a little, or by doubling it up to increase the thickness. Where wall hanging isn't practical, we've often used hefty mic stands or cheap T-bar lighting stands, but there are countless options open to anyone with minimal DIY skills. For example, you could stitch curtain rings to one edge of the duvet then hang it from a curtain rail fixed to the ceiling or across a corner.

The role of the duvet here is essentially to absorb any reflected sounds that pass behind the singer and that would otherwise reflect off the back wall straight into the front of the microphone – and that alone can make a huge difference. By extending the duvet around to the sides you can also reduce the sound reflected into the sides of the mic, with further significant benefits.

Where the ceiling is low, which is the case in many home and garage studios, you may also need an absorber above the singer and microphone to deal with reflections from that surface, although this can be as simple as an acoustic foam panel or another duvet, if you can find a means to suspend it. On more than one occasion we've fitted hooks on opposite walls or picture rails to allow us to create a 'web' of nylon cord allowing foam panels to simply be rested on top.

Of course, if the acoustic treatment you've already applied to your studio to improve the monitoring acoustics includes a large area of absorber somewhere, with enough space for the singer to stand in front of it, you might find this alone does the trick, and you may not need that duvet after all!

You may well have seen commercially-produced curved absorbers, intended to be mounted behind the microphone. These can bring about a further worthwhile improvement in the recorded sound by shielding the sides and rear of the mic, and also by absorbing some of the direct sound from the singer that would otherwise go out into the room to be reflected from other room surfaces. In our experience, these behind-the-mic absorbers are rarely adequate on their own, but certainly bring a worthwhile further improvement once you have already dealt with the rear and side wall reflections.

An improvised vocal booth, using a duvet and a behind-the-mic absorber, such as this Reflexion Filter, can often produce better results than a purpose-built set up, unless the latter is very carefully designed.

Overall, by using the V-shaped duvet corner in concert with a curved absorber behind the mic, the vocal sound you get is often much better than that which can be achieved using a small vocal booth, because unless the latter is extremely carefully designed it can very easily make vocals sound boxy and congested. Unless the acoustic treatment inside a vocal booth is very thick, the low-end and lower midrange isn't absorbed as effectively as the upper mids and highs, and these lower frequencies will therefore tend to dominate the overall sound captured by the mic. We've come across several DIY vocal booths on our travels and none of them sounded good without undergoing significant modification, and in some instances there just wasn't enough physical space to apply any practical acoustic fixes so the only option was to rebuild or abandon them.

CASE STUDY – BACK TO FRONT

Always check the spec to see which part of the mic you should sing into. Most large diaphragm studio capacitor mics are so-called 'side-fire' designs, which means you sing into the side of them, whereas in contrast, most stage vocal mics are 'end-fire' mics where you sing directly into the end. (These side- and end-fire terms are nonsensical, of course: nothing comes out of them; sound only goes in, but they are widely used, nonetheless!). Usually the 'hot', or active side of a side-address microphone is designated by the manufacturer's logo, or in the case of Røde, a gold spot – we mention the latter only because on one *Studio SOS* visit we noticed that the user had a 'side-fire' mic set up in an 'end-fire' configuration, with a pop shield over the end of it.

When we asked why, he said it was because he'd, "tried using the side, but it sounded very 'roomy'… and that singing into the end seemed a lot better". Of course, what had happened was that he'd initially tried the back of the mic which, being a cardioid-pattern model, would indeed sound dull and very roomy indeed, but then rather than check the other side, he'd assumed that it was an 'end-fire' mic, and rigged it accordingly. Singing into the top of the mic meant he was actually using it 90 degrees off-axis, where the output level is only about 6dB lower than the on-axis level, and only slightly duller than it should have been! That was clearly a massive improvement on using it directly from the back in the cardioid null, but still not as good as it was capable of!

If in doubt about the polar pattern or the orientation, listen to the output of the mic while moving around it in a complete circle talking or singing as you go. That will quickly reveal the polar pattern and intended main axis!

Position

Because of the acoustic anomalies that often occur in the exact centre of a square or rectangular room, it is best to avoid placing the singer and mic in the centre, but at the same time you don't want them too close to walls either, and especially not in a corner, as the bass-end builds in intensity close to boundaries. Also, if there is a computer in the room, try to arrange things so your vocalist is set up as far away as possible, with the rear of the mic facing the computer, in order to minimise the amount of fan noise picked up.

Pop shields should always be used, regardless of the type of microphone or any claims made by its manufacturer as to its inherent resistance to popping. An inexpensive mesh, foam, or perforated-metal pop screen, placed one to two inches in front of the mic capsule, will prevent unwanted pops or thumps whenever loud 'plosives' – 'P' and 'B' sounds – are sung. What happens is that the very strong air blasts strike the mic capsule's diaphragm with such energy that they cause it to 'hit the end stops' generating a huge and unwanted low-frequency output that is very hard to remove after the event. Of course, it's best if the vocalist develops the technique of turning their head slightly when singing those plosives, in order to avoid sending strong blasts of air straight towards the mic in the first place, but few untrained singers have the ability to do that without it getting in the way of their performance. Engaging the low-cut switch on the mic or preamp, where available, will help

> It's always worth using a pop shield for vocal recording. It shouldn't compromise the recorded sound at all, but a destructive 'plosive' in the recording can't really be fixed by subsequent processing.

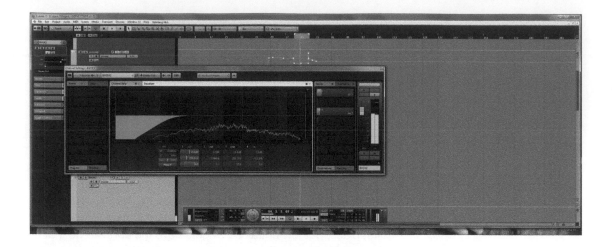

slightly, but you need to be sure that it doesn't compromise the recorded vocal tonality.

If you find yourself desperately in need of a pop screen without having one to hand, you can always improvise one using a couple of layers of fine nylon stocking material stretched over an opened out wire coat hanger. We've also used the fine-wire mesh splashguards designed to fit over frying pans, with good results – you just need to improvise a means of fixing them in place. The inherent nature of a pop shield is that most tend to attenuate the very highest frequencies to some degree, with multi-layer nylon screens being the worst and open-cell foam being the best. However, any losses can be made up easily with a little 'air EQ' in the mix, and this minor drawback is far more acceptable than the huge plosive pops that would occur without a pop shield in place!

The distance between the mic and the singer should normally be about 200 to 300mm – or slightly further if the mic has a strong proximity effect – with the pop-shield about 100mm from the mic. It often helps to set the mic capsule around forehead height, so the singer has to project slightly up towards it. This sometimes achieves an improved posture that may help them sing more effectively, but it also helps to make sure plosive blasts go below the mic rather than straight into it, and also tends to pick up less sibilance.

TIP: Even a good vocal recording made with a pop shield will show some unwanted low-frequency information when you look at the recording using a spectrum analyser. That's because some air blasts from the singer's mouth still get through the pop shield, albeit heavily attenuated. Using a low-cut, 18dB/octave filter, set somewhere between 80 to 200Hz, will address this, but always listen to the audio as you're adjusting the filter to ensure that you don't unduly affect the intentional low end of the voice.

With many singers, it helps to set the capsule slightly higher than their mouth height, making them project slightly upwards.

Mic Choices

We said at the outset that you didn't necessarily need an expensive mic to get a good vocal sound, but there is an important *caveat* to add: the critical aspect is to use a microphone with a tonality that fundamentally complements the voice of the singer you are recording. For example, some vocal microphones are quite bright and 'airy', some have a strong 'presence' character, and others are warm-sounding to the point almost of being dull. All mics have a tonal character of some kind, even the 'neutral' ones. One of the first principles in achieving a good vocal sound is to find a mic that doesn't require much (or indeed any) EQ to make the singer's voice sound natural. It sometimes helps if the mic's fundamental characteristic tends to lean in the opposite direction to the singer's own vocal character – a warm-sounding mic for a singer with a thin, shrill voice, whilst someone with a warm voice, perhaps lacking in definition, might pair up well with a bright-sounding microphone.

If you can only afford one microphone, and you need that model to work with a number of different singers, then try to pick something fairly neutral, as you have more chance of being able to use a modest amount of EQ to get the sound close to where you want it. In general we have found that finding a suitable microphone for a male vocalist is relatively easy – most mics seem to give acceptable results in most cases. With female voices, however, it seems to be far more difficult to find the ideal combination, possibly because the 'presence peak' built into the response of many mics tends to emphasise a naturally more resonant area in female voices. In the various shootout tests we've done, the suitability of a mic to a particular voice really has had little correlation with its price. Experimentation is the key.

'Comfort Reverb'

Those may be the practicalities of setting up for vocal recording, but no amount of technology will help if your singer doesn't turn in a good performance. To maximise your chances of achieving this you have to work with them to set up the best possible headphone mix with exactly the right balance of the backing track and their own voice, as well as just the right amount of 'comfort reverb' for them to feel confident. We have found that this is an area that's often neglected in home studios with some users even being unsure as to how to set this up.

When working with DAW software-based setup, there are two general scenarios: firstly, using the DAW's 'software monitoring' facility so you hear the voice after it has passed through the DAW's processing, usually with a tiny delay; secondly, using the 'direct monitoring' facility often built into audio interfaces – this allows you to hear the input signal before it passes through the software. Using a hardware mixer in conjunction with the interface, as many people do, is effectively just another way of achieving direct monitoring.

If you are using software monitoring, the perceived delay is usually manageable with buffer sizes up to 128, although some singers and instrumentalists are far less tolerant of this than others. The problem is that they hear their own voice immediately through internal bone conduction, followed by their voice arriving back in their headphones a couple of milliseconds later (after passing through the mic, interface, computer and interface again). The delayed signal interacts with the direct signal to cause colourations that can be surprisingly off-putting for some, but completely innocuous to others!

Setting up a comfort reverb using software monitoring of the live input in the computer is very easy, as you can simply set up an aux send routed to a reverb plug-in from the input channel itself. You can choose whatever type and amount of reverb makes the singer comfortable, as it won't be recorded with the track.

Achieving the same when using direct monitoring via hardware is slightly more tricky, and I've always found the easiest way to do it is to leave software monitoring on in the DAW and then set up a pre-fade aux send on the track you're recording to, feeding this to an aux bus with a suitable reverb inserted.

Pre-fade send used for temporary 'comfort reverb'

Reverb plug-in (fed pre-fade) returned to main output

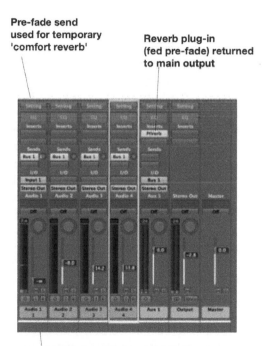

Track being recorded has fader fully down so only the direct-monitored sound of the input is heard, along with the reverb return

Configuration for achieving 'comfort reverb' from your DAW in the headphone feed, whilst using latency-free, direct input monitoring via the hardware interface.

Interface monitor control sets DAW/Direct balance

The channel fader can then be turned right off to avoid you hearing the voice slightly delayed 'through the DAW' – you will get the direct voice from the direct monitoring path in the interface. Since the aux send is pre-fader, though, you can still send some vocal to your comfort reverb without the processing delay being noticeable – effectively, a few milliseconds will just be added to the reverb's pre-delay parameter. Again, use whatever type and amount of reverb makes the singer comfortable.

If this comfort reverb facility is a feature you're likely to use regularly, it is well worth setting up the necessary pre-fade send and a suitable comfort reverb as part of your template song. It always surprises me when we visit readers' studios how few of them work from a template – instead they'll waste a lot of time setting up the tracks, busses and plug-ins over and over again for each new song they start. Templates can save a lot of time and you can always update them when you discover new things you're likely to need on a regular basis.

The Studio 'Vibe'

Even in a bedroom or garage studio, you can do a lot to make the singer feel more comfortable – even when the singer is you! Ensure sure the room is a comfortable temperature and always have a glass of water (at room temperature) on hand. Turn off the lights or use subdued table lamps if that helps create a better atmosphere, and if a guest singer is still feeling self-conscious, send all unnecessary musicians out of the studio until you're done. Always be positive and encouraging, so rather than telling them they've done a dreadful take, just say you think they've got an even better one in them! Always record everything, too, even the warm-up, as some of the best moments sometimes occur then. As a rule I'll record three or four takes of the whole song, then compile a 'best of' version from the best phrases. If any further replacements are necessary, then I'll resort to recording individual phrases, offering advice on performance where necessary. We now have software that can correct small tuning discrepancies pretty effectively, but the feel of the performance still needs to be there.

Artist Monitoring

While recording, vocalists always need to hear the recorded backing parts, of course, and this is generally best accomplished via headphones. It is usually sensible create a conventional balance of everything as a starting point, and then adapt the mix as the vocalist directs, remembering that it is vitally important to have clear backing instruments that guide the vocalist's pitching.

For singers who prefer to work with one side of the headphones pushed back so they can hear a little of their voice directly, it is important to make sure that the unused

Some singers prefer a thinned-out mix in their headphones when overdubbing vocals, but you need to make sure there are still enough cues for accurate timing and pitching.

Headphone 'spill' can have a seriously
detrimental effect on the tonality of your
recorded track, especially if the singer
prefers to only use one earpiece. Make
sure your performer keeps the unused
earpiece pressed tightly against them, or
change your foldback routing to only feed
the other side.

side of the phones is kept snug against the side of their
head to prevent sound leaking into the microphone. If they
can't do this ('big hair' can be a problem) it might even be
necessary to route the cue track only to the side that is
actually being used.

Some vocalists are simply never comfortable singing with
headphones on, however. It is actually perfectly possible to use
loudspeaker playback whilst recording vocals, particularly for
louder musical styles where a relentless backing track will tend
to mask the additional spill that inevitably results. Switching to
a hand-held, stage-type, vocal microphone, designed to be
used close-up, helps to minimise spill from speaker playback
and keep it to manageable levels, and make sure the playback
speaker is aimed at the null of the mic's polar pattern. If you
want to get really ambitious with speaker monitoring and still be
able to use a proper studio mic, there is a technique involving
using two speakers fed the same (mono) signal, but in opposite
polarities, placed on either side of the microphone. With
careful positioning, it is usually possible to achieve substantial
cancellation of the playback signal as it is 'heard' by the
microphone, whilst the singer can hear the track perfectly, due
to the spacing between the ears and the shadowing effect of
the head.

Two speakers carrying the same mono signal, but with the polarity reversed on one of them, can be rigged on opposite sides of the mic, at exactly the same distance to create a headphone-free overdub setup. The out-of-phase speaker signals cancel out at the microphone, but are audible to the singer due to the distance between the ears and the shadowing effect of the head.

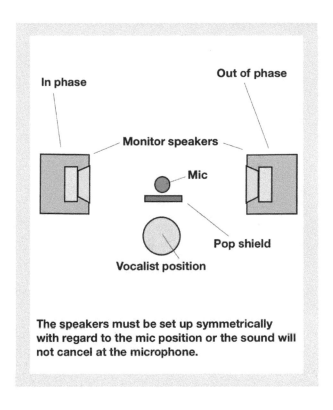

In phase

Out of phase

Monitor speakers

Mic

Pop shield

Vocalist position

The speakers must be set up symmetrically with regard to the mic position or the sound will not cancel at the microphone.

TIP: When you do your initial sound-level check, leave at least 12dB of headroom (and maybe as much as 18dB) because the singer will almost certainly turn in a louder performance when the red light goes on. On so many of our *Studio SOS* visits we've found DAW mixes with individual tracks peaking so high that the stereo bus is clipping. There's no significant noise or resolution penalty for leaving a generous amount of headroom on today's DAW systems, so we'd recommend leaving at least 12dB for whatever source you're recording. No matter what the DAW designers tell you about the 'floating-point cleverness' of their mixers, we've always found that mixes sound cleaner and less fatiguing when you leave plenty of headroom.

Processing Prior to Recording

Many professional engineers use compression both during recording and then again when mixing. Compressing the signal while recording increases its average level, and that was an important consideration when recording to analogue tape or early 16-bit digital systems. Now that we have 24-bit recording, however, there's no technical reason to compress during recording at all unless you have a hardware compressor that imparts a particularly desirable quality to the sound. Even if you have an analogue compressor that sounds very special, it's usually best to err on the side of applying less compression than you think you may ultimately need – it is very easy to compress more when you're mixing if there isn't enough, but virtually impossible to undo the effects of over-compression. When working on *Mix Rescue* projects for *Sound On Sound* we often come across tracks that have been over-processed at

source, and this makes mixing very difficult. If in doubt, record vocals and other instruments completely clean and flat – no compression, no EQ. If the sound isn't close to what you want, then change the mic position or the microphone, and get it sounding as close as possible to the way you want it before reaching for the EQ.

Summary

The most important areas to attend to when making a vocal recording are the acoustics of the recording space and the comfort of the artist themselves. Neither of these is particularly budget-intensive: a duvet or two, plus a few minutes setting up a good monitor mix with the right amount of comfort reverb, will make more difference to the end result than upgrading your mics and preamps.

Chapter Eight
Acoustic Guitars

:::: ### Recording acoustic guitars

After vocals, the acoustic guitar is probably the next most common musical source in the typical project studio to be recorded using a microphone. Many acoustic guitars come with a factory-fitted piezo-based undersaddle pickup and

preamp, and whilst these can be effective in a live performance environment, they don't offer anything like the same complex tonality you can achieve by miking the real thing. There are dedicated acoustic guitar preamps that use digital modelling to achieve a more realistic representation of the complete sound of the guitar, and they can work well in the context of a mix if you are in a situation where spill precludes using microphones, but where possible we'd always recommend that you use microphones for this particular source.

In the course of our *Studio SOS* ventures we've encountered many acoustic guitar recordings that could easily have been improved, with typical problems ranging from basic tuning issues to poor choice of microphone or mic position. Provided that the strings are reasonably new, it should always be possible to get a usable recorded sound from all but the very cheapest acoustic guitars. So long as the action is adequately low for comfortable playing, without any significant fret buzz, you probably don't need to go to the expense of a full professional setup. If you are competent with these things and have the necessary tools, it might be worth ensuring that your guitar's truss rod is set to keep the neck just short of perfectly flat (just marginally concave) as

◄

The rather one-dimensional sound typical of an under-saddle, piezo-based acoustic-guitar pickup can now be transformed into something rather more realistic by the sophisticated digital EQ and phase manipulation offered by this Fishman Aura pedal.

this usually helps to minimise buzzing and choking, as the vibrating strings have a tiny bit more excursion room.

A common reason why a recording of a good guitar may sound less great than it should, even with a well-chosen mic and suitable mic position, is that the recording environment is too dead. We've heard this on many occasions where the guitar has been recorded in a carpeted bedroom surrounded by lots of soft furnishings. Having a fairly dead acoustic environment works well for vocals, but in the case of acoustic instruments some reflections from hard surfaces can help breathe life into the sound.

Acoustic guitars usually sound best when played above a hard, acoustically reflective floor surface, such as tiling, solid or laminate wood, or linoleum, as this brightens the sound. In situations where people have had to record in carpeted rooms we've been able to demonstrate that putting a modestly-sized sheet of MDF or hardboard on the floor between the guitar and the mic works almost as well as having a room with a hard floor throughout.

> Acoustic guitar recording benefits from controlling room colouration just as much as a vocal recording.

As when recording vocals, it is usually best to keep the guitar and microphone away from the walls and also away from the exact centre of the room. Exactly the same acoustic treatment strategies for killing unwanted reflections can be used when recording acoustic guitar as you might when recording vocals – duvets, acoustic foam, heavy blankets and so on. Where a reflective floor can't be improvised, siting the instrument between the centre of the room and one reflective wall can also help the tone, but as a rule try to avoid multiple reflective paths, as that can lead to a coloured, 'roomy' sound. One reflective surface close to the instrument usually works best. Furthermore, as explained in earlier chapters, try different places in the room as the room's modal resonances can have a profound effect on the recorded result, especially at the bass and low-mid end of the spectrum.

As the acoustic guitar covers a wide frequency range with a strong transient content, the best recording results are usually achieved using capacitor microphones. It is no surprise then that many of the less satisfactory results we've come across have been due to the use of dynamic vocal microphones but, with care, even these can yield passable results with appropriate placement and some EQ. Small-diaphragm capacitor models are the first choice for most acoustic instruments because of their excellent transient response and extended high-frequency capture, but you can also get good

◄

Small-diaphragm capacitor mics are often favoured for acoustic instruments because of their excellent transient performance and extended high-frequency response.

results with a large-diaphragm model too. Some home studios may only have one really good mic, and the chances are that will be a large-diaphragm capacitor model, usually with a fixed cardioid pickup pattern.

It is also possible to get a very nice, natural sound using a ribbon mic, although you may need to employ EQ to tame the inherent low-end boost due to proximity effect, and perhaps enhance the highs a little, in some cases. Ribbon mics have a fairly low output and so require quiet preamps with plenty of gain, so the kind of preamps typically found in budget audio interfaces might struggle when used with a ribbon mic on a relatively quiet source like acoustic guitar.

Although we tend to think mainly of cardioid-pattern mics for project-studio recording, a microphone with an omnidirectional pattern often produces the nicest sound for acoustic guitar – although you will probably need to put absorbers around the mic to cut down on the capture of unwanted room reflections other than from the floor. Omni mics don't have the off-axis colouration or the low-frequency phase shifts that are inherent in most cardioid microphones, and so tend to produce a more natural sound. An omni-pattern mic picks up sound equally well from all directions so

> Ribbon mics require quiet preamps with plenty of gain. Cloud Microphone's Cloud Lifter can significantly improve the performance of the typical preamps found in budget audio interfaces when using any type of low-output moving coil or ribbon microphone.

mic positioning with respect to the instrument is actually less critical than when using a cardioid-pattern mic, and there's also no proximity effect to cause unwanted bass boost. With such a physically wide sound source as a guitar body, pretty much all of which vibrates, much of the sound will be occurring away from the main axis of the microphone, so an omni model clearly has the best chance of capturing this more accurately than a cardioid pattern mic.

Figure-of-Eight Microphones

It is rare that home studio owners reach for a figure-of-eight microphone, as the fact that they are equally sensitive front and back tends to make them worry about spill or picking up excessive room sound. However, we've demonstrated on more than one occasion that a figure-of-eight microphone can be a real life-saver when you have a performer who, in order to get the right performance, needs to sing and play guitar at the same time. Miking both sources with cardioids rarely achieves sufficient isolation between the guitar and the voice to allow you to optimise both of them. The reason is that the unwanted sound source arrives at around 90 degrees off-axis, and the cardioid pattern is only about 6dB less sensitive at that angle than it is directly on-axis. So the vocal mic inherently picks up a lot of guitar sound, and vice versa – but if you're using large diaphragm mics that off-axis sound will also be quite coloured and unnatural-sounding.

The real strength of the figure-of-eight mic in this situation is it is has a 'deaf' null at 90 degrees off-axis on both sides. If you can arrange the guitar mic so that its deaf axis points towards the singer's mouth and, similarly, that the singer's mic has its dead angle aimed towards the guitar body, you can claw back quite a lot of separation. Of course, it is important that the performer stays quite still and doesn't sway about, to avoid moving the unwanted sources away from the null axes. This arrangement may also require more acoustic screening to prevent unwanted sound getting into the rear of the mics, but sometimes it presents the only practical solution. Note, though, that unless you intend to use pitch correction software or heavy effects on the voice

Figure-of-eight mics have a 'deaf' null at 90 degrees off-axis on both sides. Arranged carefully, they can offer a worthwhile amount of separation even on sources that are quite close together, such as someone singing and playing guitar simultaneously.

TIP: The commercially available curved sound absorbers designed to be used behind vocal mics are also very effective if placed behind omni or figure-of-eight mics when recording acoustic instruments, as they cut down on the unwanted room reflections reaching the rear of the mic. Rather like an omni, most figure-of-eight mics don't suffer tonal colouration as you move off-axis, though the overall level falls away to zero once you get to 90 degrees off-axis. However, you do have to take into account the fact that figure-of-eight mics have a strong proximity effect, which means you either have to position them further from the source or employ a low-cut filter to balance the sound.

(or guitar), you don't need to achieve perfect separation – just enough to allow you to balance and, if necessary, EQ the two sounds adequately.

Finding the Optimum Mic Position

When trying to record an acoustic guitar, the first challenge is to locate the best position for the mic. Usually the optimum microphone distance is between 10 and 18 inches (25–45cm) from the guitar body where the mic will 'hear' a large part of the guitar's vibrating surfaces, and there will be little proximity effect if using a directional mic. A mistake commonly made by inexperienced musicians is to place the mic directly in front of the guitar's sound-hole. It may seem like this is where most of the sound comes from, but unless the guitar has a very small body this position invariably produces an unpleasantly muddy and boomy tone with poor definition. A much better starting point is to direct the microphone towards the point where the guitar's neck joins the body, as this usually produces a much better-balanced sound. However, every guitar is different and while this mic position will probably provide a very acceptable sound, don't automatically assume this will give you the *best* or most representative sound for the specific guitar. One of the things we often do during *Studio SOS* visits is to ask the client to play the guitar while monitoring the sound via headphones. As they play we move the mic around so that they can hear the effect of changing the mic position, and in most cases they're absolutely astounded at how dramatic a difference this makes.

Where somebody else is playing the guitar, we have found that the best method for finding the optimum mic position is simply to monitor the microphone's output using good quality headphones, and while the musician plays you can then move the mic around by hand until you find a position that delivers the best sound. With the ideal positioning identified you can then fix the mic on a stand. What you're listening for is a tone that's both warm and lively, but without being harsh or boomy and without any unnatural phasiness. Two common mic position alternatives to the neck/body joint, that often work well, are firstly below the guitar body looking back up towards the bridge – which tends to give a more rounded lower end weight with many guitars (and less fret noise) – or to place the mic alongside the player's right or left ear looking down over their shoulder. If you want a thinner tone, perhaps to help the guitar cut through without muddying the mix, then moving the mic more towards the headstock end often helps.

Provided that you take care with the microphone placement at the recording stage, you should be able to capture a natural sound that works well in the mix without requiring much further processing. However, unless the recording is of just a solo guitar, some equalisation often helps to make it blend with other components in the mix. Where a strummed acoustic guitar is part of a busy arrangement alongside bass and drums, filtering the low end quite severely – reducing everything below maybe 300Hz using an 18dB/octave low-cut filter – is standard practice, as this produces a guitar sound with very little body but plenty of percussive definition. Don't worry about how the guitar sounds in isolation when soloed – it will obviously sound

Finding the 'sweet spot' is much easier if you are monitoring the output as you move the mic around.

quite thin – It's how it works in the context of the complete mix that's important. If used as a rhythmic bed, the equalised acoustic guitar should almost blend with the hi-hat sound.

If you feel the guitar needs more clarity and presence, a shelving-EQ boost in the 7 to 10kHz region will add sparkle and polish to the result without reaching far enough down the spectrum to give the sound an aggressive edge, but once again you must evaluate the results by listening to the complete mix rather than just the guitar in isolation. Problems we've encountered often involve the user boosting at too low a frequency, typically in the 2 to 4kHz range which, although it helps the guitar cut through in a mix, can also make it sound unpleasantly harsh. Boominess or boxiness can be tackled using a suitably tuned parametric equaliser in 'cut' mode, if a simple, shelving low-cut doesn't do the trick – but it's always better to avoid this problem in the first place by taking the time to find the best mic position that avoids capturing a boomy sound. If you need to cut out boominess using EQ, start by

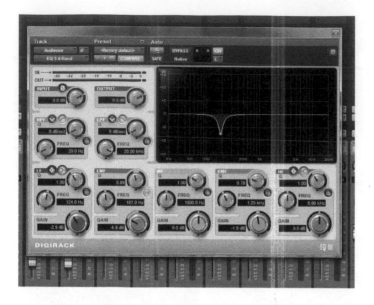

◄ Narrow-band EQ cut can help tackle a single boomy or boxy-sounding frequency without altering the overall tonality.

sweeping through the 100Hz to 300Hz range with a moderately narrow-bandwidth EQ boost, in order to locate the trouble spot. When you've found the unpleasant boomy frequency adjust the gain control to apply as much cut at that frequency as is necessary, while also tweaking the bandwidth or Q control to affect only the troublesome frequency range.

Stereo Recording

It might be tempting to record the acoustic guitar in stereo, and for solo performances this approach can work well. However, as with any multi-mic technique you can run into phase-cancellation problems if a spaced mic technique is employed and the stereo mix is later summed to mono, so always double-check the stereo miking using the mono button in the monitor controller or software. You can use any of the standard spaced or coincident (including MS) stereo mic techniques if you wish, but a lot of engineers take a more pragmatic approach and settle for two physically unrelated mic positions and then pan the microphones to achieve a musically pleasing spacious result, rather than a theoretically 'accurate' stereo image. A common ploy is to mic the guitar's body, as

TIP: Ultimately every sound has to be judged in the context of the whole mix. It is best not to spend too much time trying to equalise individual instruments to sound as close as possible to the 'finished' sound you think they will need in the mix, whilst recording them. You may well have to change your approach once the rest of the track is playing and if you have used too much EQ or compression, you may have already compromised your options.

described earlier, to get a rich and full tone, combined with a second mic part way up the neck to create a brighter, thinner tone. Although this obviously doesn't give a true stereo picture of the instrument, it often works well in practice. The main pitfall we encounter when the guitar is recorded using spaced microphones is that the stereo image shifts as the performer moves, even if the movement is only slight.

Where the recording environment has a very attractive acoustic of its own, a more traditional coincident or near-spaced mic technique might be appropriate to place the guitar within that acoustic in a natural way that captures the spaciousness of the room in stereo. The coincident XY or Mid-Side techniques usually work well in this context, as does the ORTF arrangement (cardioid mics spaced 17cm apart and angled at ±55 degrees). In this scenario the mic distance from the instrument needs to be adjusted in order to pick up just the right amount of room sound.

Mic positions are often chosen to try to give a complete picture of the tonality of the guitar, rather than a spatially accurate stereo image – in this instance, the body mic gives fullness, whilst the neck mic adds articulation.

In most cases, however, we've found that the guitar presents a more stable sound source if recorded in mono, as this avoids any phase-related mono-compatibility issues and image-stability complications. You can add a sense of 'space' later, if desired, by using a suitable stereo reverb. Compared to the dimensions of the listening space, the acoustic guitar is a relatively small sound source, so at the typical listening distance of an audience there is no significant source sound width anyway. Any stereo information you hear is coming from room reflections, and these can be emulated perfectly adequately using a suitable reverb – whether a hardware unit or a DAW plug-in. This 'keep it simple' mantra crops up often, as many of the problems we discover during *Studio SOS* visits are due to a process being made more complicated than is strictly necessary.

Improving DI'd Guitar

Sometimes there is just no practical alternative to DI'ing the guitar (recording the output of the pickup directly) – such as when it is being recorded at the same time as other, louder instruments, or where adequate separation can't be achieved when using a microphone. Some specialised acoustic guitar preamps attempt to put back what the DI process leaves out – specifically the complex body resonances of the instrument – using digital modelling, but if you don't have access to one of these products, there's an alternative that is well worth trying.

Many DAWs now include a 'Fingerprint EQ' or 'Match EQ' plug-in, but if your particular DAW doesn't have one there are several third-party versions available. Essentially the fingerprint EQ combines a multi-band equaliser with a spectrum analyser, allowing you to analyse a recording of the sound you want to achieve – the 'target' sound – and then use that data to automatically EQ your source sound to achieve a similar spectral balance. It is a process that can be a bit hit and miss, especially when used on complete mixes, but it can produce very good results and is always well worth trying. You can use someone else's recording for your target sound, of course, but the best results usually come from miking the actual guitar you

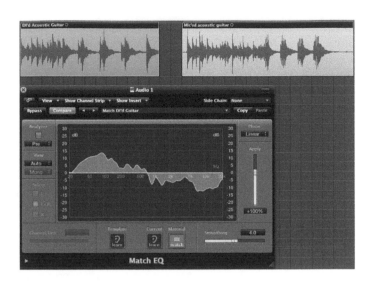

'Fingerprint EQ' imposing the response shape of a miked instrument onto a DI'd pickup signal.

are using and playing the same chords as you will be using in the recorded DI part, as the lumps and bumps in the frequency response are key-related, to some extent. If the calculated EQ doesn't sound quite right you can sometimes get better results by manually applying a similar, but simpler, EQ shape using an ordinary EQ plug-in instead.

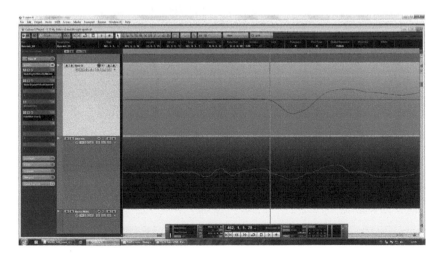

Simultaneously-recorded mic and DI signals from any instrument will often benefit from being pulled into time alignment – the DI signal will always have an earlier arrival time than the mic.

We've also had good results using both the miked and DI'd output from an acoustic guitar, panning both signals a little left and right of centre to create the illusion of stereo spread. Where spill was previously a problem, this may allow it to be brought under control without losing the contribution from the microphone entirely. If you try this dual approach, it is worth zooming in on the two audio waveforms to ensure that they are in exact alignment as the mic signal will be slightly behind the DI signal and the timing difference can compromise the tonality. You may also find that one of the waveforms is inverted in polarity relative to the other, in which case one of them needs to be inverted again to bring it back in phase. Where there is a timing discrepancy, the miked track can be advanced in time slightly so that it aligns with the DI'd version.

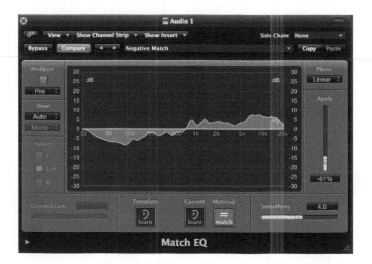

∧

TIP: An unlikely 'improver' trick – most fingerprint EQs have an 'amount' slider that lets you adjust the degree to which the source sound's spectrum is modified to match that of the target sound, and in some cases the amount slider will include a 'negative' capability so that you can make sounds more dissimilar. The initial idea was to use a section of vocal as the target sound, then to apply this setting to the guitar sound to try to achieve more separation by moving the fader into its negative area. Used this way, it works, in theory at least, to carve vocal-shaped holes in the guitar spectrum. We wanted to know if this would help make clearer-sounding mixes while still keeping the guitar sounding reasonably loud. It did work to a degree, but more importantly the complex peaks and troughs in a typical vocal spectrum, when applied in reverse to the guitar sound, seem to add a surprising degree of complexity to the sound and can help make a rather bland acoustic guitar part sound as though it was played on a better instrument. We can't guarantee this will always work, as it depends both on the original guitar sound and on the vocal spectrum used as a target, but it is certainly worth experimenting with the idea.

Chapter Nine
Electric Guitar and Bass

One thing that has surprised us during our many *Studio SOS* visits is the number of home studio owners who play guitar as their primary instrument. It is often assumed that keyboard players are the ones who take most naturally to computer recording techniques, mainly because they were at the leading edge of the widespread adoption of MIDI, which revolutionised music production from the mid-80s onwards, but we've come

The electric guitar is perhaps unique in that the amplifier and speaker cabinet it is connected to are just as important as the instrument itself in influencing the final sound.

across home studios with high powered Marshall stacks set up in cupboards underneath the stairs, innumerable Line 6 POD processors, and even a band trying to record two large guitar stacks and a bass amp in a small garden shed alongside a miked drum kit! It is probably fair to say that there are more ways of recording the electric guitar than just about any other instrument.

The electric guitar is perhaps unique in that the amplifier and speaker cabinet it is connected to are just as important as the instrument itself when it comes to creating the final sound. Take Leo Fender's wonderful Stratocaster guitar design and give it to a bunch of different players, with different amplifiers – or even the same amps with different settings – and you get sounds as diverse as those made by Jimi Hendrix, Hank Marvin of the Shadows, Mark Knopfler of Dire Straits, Dave Gilmour of Pink Floyd, Eric Clapton and countless others. Since the amplifier is so important, how you choose to record the electric guitar is an important creative decision, yet so many people just use whatever amp-modelling plug-in they have to hand in their DAW simply because it is convenient. That's not to say you can't get good results this way, but so much depends on the style of music and on the player.

Model Behaviour

Digital-modelling preamps and their software plug-in equivalents vary in quality, with some sounding very close to the amps they purport to model, but to our ears none manage

◄

Digital guitar-amp modelling devices offer an easy and effective route to a wide variety of recorded guitar tones for the small home-studio owner who may not be able to record a real amplifier, perhaps due to noise constraints.

to fully convey the essence of a high-energy rock sound in the same way as a miked amp. Part of the reason for this is that when you stand in front of a guitar amplifier some of the sound energy from the speakers interacts with the strings and guitar body to help sustain and energise the sound. Despite a number of electronic and software-based sustain innovations, we think it is fair to say that nobody has found a way to emulate this complex interaction with any degree of authenticity. For many less assertive styles, however, you can get a very acceptable sound using a modelling preamp or plug-in, and in those cases all you have to do is set the record level, leaving a sensible amount of headroom, as always, and then go for it.

If you happen to have both a hardware modelling preamp and a recording system that offers an amp modelling plug-in, then try recording via the hardware unit using a little less distortion than you think you'll ultimately need – or even split the guitar signal so that you can also record a clean version that you can process later using plug-ins. You can always add more

TIP: If you're using a modelling amp plug-in you can often get better overdrive sounds if you use a real hardware overdrive pedal connected between the guitar and the input to the recording system, rather than using a software model of an overdrive pedal. The hardware pedal feeds into your virtual guitar amp plug-in, just as it would with a real amp. This technique has a further advantage in that the pedal output can be plugged into a regular line input in your recording interface if you don't have a high impedance DI box or instrument input available. This works because most pedals don't mind driving the relatively low line-input impedance, whereas an electric guitar's pickups would object with a low level and a compromised tone. Just be aware that there's no point in cranking the pedal's output up to drive the input harder, as you might with a real valve amp, when the pedal is feeding a solid-state, line-level input. In this context, you are using the pedal for tonal shaping and a bit of compression and distortion, rather than achieving overdrive through level boosting.

distortion and make further tonal changes by processing the recorded parts via your amp modelling plug-ins when you come to balance the mix. You can also add further sustain to a recorded guitar part by using compression, rather than more overdrive, a technique that can keep a mix sounding lively while avoiding the muddy congestion that layering too many heavily distorted guitar parts can cause.

The latter is a surprisingly common problem: guitar parts recorded with far too much distortion can make the part sound indistinct and muddy in the context of a mix, and this is something that is very hard if not impossible to redress with signal processing. We occasionally also find parts recorded so high in level that the recording itself clips in places, and this can lead to a harsh and unpleasant sound that is impossible to rescue or disguise adequately, even using a distorted-amp plug-in.

Miking a Guitar Amp

A surprising number of classic rock albums achieved their big guitar sound through the use of small tube amplifiers driven hard – a methodology that is well suited to the home studio, as long as noise leakage isn't an issue. Many of the classic Led Zeppelin albums were recorded using a small practice amp, while Eric Clapton's *Layla* album was recorded using a very low-power Fender combo with a small single speaker. It is much easier to get the right recorded sound at a manageable volume this way than with a big stack – although if you have enough space and tolerant neighbours a big stack can sound great too!

A typical scenario we encounter during our *Studio SOS* visits is that the operator hasn't taken the time to experiment with the microphone type and position to achieve the best sound, assuming that the required sound can be found by processing when mixing. The reality is that the degree to which an already overdriven electric guitar sound can be knocked into shape using EQ and other plug-ins is more limited than you might imagine, so it pays to try to get close to the desired final sound at the miking stage.

The usual technique with a 4x12 is to first locate the best-sounding speaker in the cabinet and then mic only that one. Effectively, the rest of the speakers are just there to annoy the neighbours, hence our preference for working with smaller combos! In a very large, good-sounding studio, a big stack might be miked up using three or more microphones at the same time: one close, one a few feet away, and one (or a stereo pair) at the back of the room. Additionally, the engineer may close-mic two of the speakers and record them to separate tracks as they may sound usefully different. These two signals might be combined in various ways to give the mix engineer plenty of options in finding the most appropriate sound. Realistically though, this approach is impractical in most small project studios where one close-mic and maybe a second at a greater distance is about as sophisticated as it gets.

Keep it Down, Lads!

Even if you do have the space and a choice of mics to get a great sound from a stack, the chances are that it will only produce the rock sound you're after at a level where the rest

A 'power soak' connects between the power amplifier output and the speaker cabinet, allowing the amplifier to be run at a high-power setting to generate distortion and compression, whilst greatly reducing the power passed to the speakers. The rest of the power is dissipated as heat within the power soak unit's high-power, wire-wound resistors.

of the street can also hear it. If you really like the sound of your stack, or just don't have a suitable small amplifier, there is always the option of using a 'power soak'. These consist of some high-power, wire-wound resistors that connect between the power amplifier output and the speaker cabinet, allowing the amplifier to be run at a high-power setting, whilst greatly reducing the power passed to the speakers. Most of these power soak units have switchable attenuation settings enabling you to feed just a few Watts to the speaker when the amplifier is running flat out. The rest of the power is dissipated as heat within the power soak unit.

/ CASE STUDY – ROCK BAND IN THE GARDEN SHED!

The following example came about as a request for help from a band recording rock music in a garden shed just about large enough to accommodate all of the band members at the same time! Noise leakage wasn't a problem, as they were well away from any neighbours and they were already using power soaks for recording two large, closed-back guitar stacks and a bass amp, alongside a miked drum kit. Their problem was that, even with the power soaks, too much of the guitar sound was getting into the drum overhead mics, making their tracks difficult to mix.

We improved the spill situation by keeping the power soaks on the guitar amps but turned their speaker cabinets around to face directly into the mattress-lined walls of the shed, leaving just enough space in front of each cabinet to place a microphone. In this way most of the high frequency amp sound was absorbed by the mattresses instead of bouncing around the room and finding its way into the drum overheads. We also took a DI feed from the preamp output of the bass amp as that can sometimes be used in addition to, or instead of, the miked feed. The band was still able to monitor their performance as loudly as they liked via headphones, but when we did our first take the amount of guitar spilling into the drum mics was too low to worry about whereas before, with the amps facing the centre of the room, the drum overheads were picking up more guitar than drums! The lesson here is that it may sometimes take a combination of two or more techniques and tricks to solve a problem. /

/ DUMMY LOADS, SPEAKER EMULATION AND BEYOND

A number of products combine a dummy speaker load with filtering circuitry that is designed to create a response similar to that of a guitar speaker. The filtering is usually adjustable to enable a few different cabinet types to be emulated. A more hi-tech take on the concept comes in the form of dummy load combined with a digital convolution-based speaker simulator. This works in a similar way to those reverb units and plug-ins that use an Impulse Response recorded in a real room or space, and will often include a library of responses taken from real guitar speakers, miked using a range of studio microphones. Some models also include power amp modelling, which allows you to take an input signal from a guitar preamp – the effects-loop send on the back of an amp, perhaps – or even from a modelling preamp.

Guitar sound is a very personal thing, so you just have to try as many options as possible until you find one that works for you. Just keep in mind that your ultimate aim is to get the right recorded guitar sound, which is not the same thing as having a fully realistic playing experience – nothing feels quite like standing in front of a loud amp!

▲
The Motherload from Sequis combines analogue, passive filtering to simulate the response of a typical guitar speaker, whilst Two Notes' Torpedo range uses digital convolution to recreate the responses of a library of real speaker/microphone combinations. Despite the different technologies, both are highly effective.

/

Amplifier Types

Today, in addition to traditional all-tube, or solid-state amplifiers we also have hybrid designs using every possible permutation of digital, tube and solid-state circuitry. While enthusiasts argue over the merits of each, the bottom line is that whatever sounds right, is right – only *you* know what you want your guitar to sound like. A practical advantage of some of the hybrid and solid-state designs is that their power stage can be turned down while the preamp (or intermediate tube stages, where used) can still be driven as hard as necessary to get the desired sound. This often means you can get exactly the sound you want at any volume you want, unlike many tube amps that have to be cranked up to a certain volume (or used with a power soak) to deliver their best sound. While some of these hybrid designs offer a 'speaker-emulated' DI output that can be used for recording, we still generally feel that the best results are achieved by miking an actual guitar speaker.

TIP: If you are using a modelling pre-amp, you may find you get a better rock guitar sound by feeding it into a small PA speaker or into the power-amp input (or effects-loop return) of a guitar combo, and then miking the speaker, rather than simply DI'ing it. This can give you back some of the interaction between the guitar and the speaker that is missing when you DI. If you feed your modelling processor into a guitar-combo power stage, though, remember to disable the speaker simulation – you don't need an emulated guitar speaker as well as a real one! However, if you're using a full-range PA speaker, be sure to leave the preamp's speaker modelling switched on.

/ SOFTWARE EMULATIONS

It's not just in hardware that guitar-amp emulation has progressed in recent years: there are now more software plug-in emulations of amplifiers, cabs and speakers than we can count. The quality of many of them is very good indeed, and the prospect of getting your computer to do all the work for a fraction of the price of equivalent hardware is enticing. To get the best out of such software, though, there are a few things you need to consider:

Monitoring Latency

If you've read this book from the beginning, you'll remember that we discussed the problems of latency in relation to monitoring a signal as you record it. While the issue is easily overcome when recording a vocal – because you can monitor the input signal and blend some reverb in with it (the delay isn't such an issue for reverbs, as it's just a little extra pre-delay) – it can, for some guitarists, be far more problematic, particularly when they need to monitor their playing via a software-emulated amp signal. Why? Well, the sound has to go into the computer, be processed via the software, and pass back out

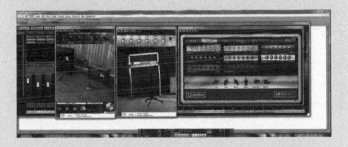

again before you can hear it and get a feel for what you're playing. Listening to a clean DI signal straight off the input just isn't the same, as you don't get any feel for the processed sound of the guitar as you're playing it – and the recording often suffers because the player isn't interacting with the sound.

Bear in mind that every six milliseconds of latency is equivalent to you being about another two metres away from the source. That's not a huge distance when you consider how far a guitarist might be from their amp on a big stage, so if you can get a good, low-latency performance from your interface this may be a non-issue for you.

Anyway, try to get the latency as low as you can while tracking by setting as low a buffer size as you can viably work with for your audio interface without introducing unwanted clicks, pops and the like. If you're planning to do overdubs later on – during a busy mix, for example – you might not be able to get as low a latency as you'd like. In these circumstances, we often find that it's better to bounce a rough backing track down as a mono or stereo mix and record alongside that in a different project, as without all that mix processing you are often able to set a much lower latency. The result can then be imported back into the main mix session.

There are other tactics you can use to get around the latency issue, though. For example, you could use a DI box to split the guitar signal, with one side going to a small practice amp for monitoring, and the other going to your audio interface's mic input for recording purposes. You might not get to hear the actual sound you'll eventually settle on in the mix, but it's much better than listening to a clean DI, and you can virtually 'reamp' the signal using your software after you've recorded it.

▲
The 'right' audio buffer size setting is always a case of finding the best compromise – smaller gives you lower latency and a faster response to controls, but larger will allow you computer to run more power-hungry plug-ins.

Try All Your Mics

Since the electric guitar doesn't have a specific 'natural' sound, you can use just about any type of microphone that comes to hand and still get something musically valid. Furthermore, because a typical guitar loudspeaker rolls off quite sharply above 4kHz or so, you don't need to choose a mic with an extended frequency response. It's perfectly possible to get a great electric guitar sound using a cheap dynamic mic with a terrible technical spec! Noise and sensitivity aren't issues either, because a guitar amp is a very loud sound source when close-miked and, in any event, the amp itself is likely to produce more hiss than just about any combination of mic and preamp! The moral here is to try every mic you have, even that cheap and nasty one you discovered at the back of the cupboard, as you never know what's going to deliver that magical sound!

a few futile minutes spent adding EQ and trying different amp models, we persuaded him to play the part again, this time using his small valve combo. We set up a back-electret capacitor mic a few inches from the speaker grille and re-recorded the part using slightly less distortion than was used on the original track. The new part sat perfectly in the mix and the player was amazed how much difference miking his little amp had made.

This again demonstrates that too often we are tempted to take the most convenient approach to a recording when that approach might compromise the end result to a very significant degree. It also underlines the reality that it is often much quicker to play or sing a substandard part again than to spend ages trying to salvage something with plug-ins and still not getting it to sound good!

Guitar Sound Styles

Taste in guitar tones has pretty much globalized these days, but it certainly used to be the case that the typical American rock guitar sound associated with many classic recordings was considered by many British players to be rather 'fizzy' or harsh. This may have been due, to some extent, on the popularity of capacitor microphones with US engineers when recording electric guitar, although it's equally possible that they simply liked a slightly different sound from British players and set their controls accordingly. The typically British rock guitar sound is slightly less brittle, and tends to be more strongly associated with the use of dynamic microphones.

Don't be afraid to experiment with combinations of microphones, and make notes of anything that works particularly well, recording the exact mic positions and level setting, so that you can recreate the setup on subsequent occasions. One classic combination miking setup for guitar recording involves both a dynamic and a capacitor model, with the dynamic mic used close up and the capacitor model placed a few feet out in front, capturing a more

The midrange emphasis offered by Shure's ubiquitous SM57 model has allowed it to remain a popular choice for guitar-amp recording throughout several decades.

complete picture of the speaker's interaction with the floor and the room.

The recent arrival on the market of a host of more affordable ribbon mics has further expanded the options for home-studio operators, because the inherent, gentle high-frequency roll-off of most ribbon mics tends to produce a smoother tonality than capacitor mics. At the same time they seem capable of delivering a more gutsy, detailed tone than most moving-coil models, some of which can sound quite nasal. Somehow ribbons are able to focus on the all the elements of the sound that you want to hear, while smoothing over the less desirable elements.

Some additional care with placement is sometimes needed, though, as ribbon mics can be quite susceptible to radiated hum fields from amplifier power transformers, while their strong proximity effect means you often need to roll off some of the low end. They are also prone to damage from strong air currents. Ribbon mics usually have a figure-of-eight pickup pattern, meaning they are equally sensitive to the rear as the front, but spill is rarely a problem in this application as they will almost always be placed fairly close to a very loud sound source. In cases where more spill attenuation is desirable, you can place an acoustic absorber behind the mic – this will also help attenuate the influence of any unwanted room resonances and reflections. Many modern ribbon mics are deliberately designed with different tonalities from the front and rear pickup lobes – one side often sounds distinctly brighter or darker than the other – so experiment with the sound from each side.

It is sensible to handle ribbon mics rather more carefully than most other microphone types, and putting them safely back into storage as soon as you've finished using them should become part of your studio routine. Never put a ribbon mic down on the floor or a desk as the strong internal magnets will draw in any metallic particles that may end up causing problems. As mentioned above, ribbon mics are prone to damage if exposed to very strong air currents or blasts – if you can feel air movement from the sound source on the back of your hand you need to think about how best to protect the mic! One solution is to put a pop shield between the mic and the speaker, although often it's sufficient to just angle the mic

▲

The inherent, gentle high-frequency roll-off of most ribbon mics tends to produce a smoother tonality than capacitor mics, whilst delivering a more detailed sound than moving-coil models.

at around 40–45 degrees to the speaker instead of facing it straight on. In this way air blasts from the cone 'brush' across the ribbon diaphragm instead of slamming it square on, and that helps to protect it from the effects of strong air currents without compromising the sound. If you need to protect the mic when not in use, but don't want to de-rig and pack it away, you can cover the mic on its stand with a polythene bag, so that any strong air currents can't hit the diaphragm directly.

Shaping the Sound via Mic Position

Mics that sound great on vocals or acoustic guitar won't necessarily be the best for electric guitar, and a budget mic that sounds terrible on just about everything else might actually give you exactly the results you want! Whatever mic

You'll get a more focused sound with strong upper mids by placing the mic near the centre of the cone, whilst moving the mic out to the edge of the speaker cone creates a warmer tonality.

you pick, it will typically be placed fairly close to the speaker grille, but we'd still suggest trying different distances to see what difference that makes to the tone. If you want to mic from further back in the room for a less 'in your face' sound, then a hard reflective floor surface will help keep the sound bright and lively. The position of the mic relative to the centre of the speaker cone also has a big influence on the tone. As a rule you'll get a more focused sound with strong upper mids if you place the mic near the centre of the cone, while moving the mic towards the edge of the speaker cone creates a warmer tonality with a different character to the high end. You can even mic the rear of an open-back combo to give more tonal possibilities.

A close-miked guitar speaker can sometimes seem to produce a disappointingly 'small' sound, compared to the sound of the speaker in the room. Under those circumstances, it is always worth augmenting the close mic with a second microphone placed a few feet from the speaker, if you have the space. As with acoustic guitars, place the amp on a hard floor or piece of board to create some reflections. This combined close-far mic technique will inevitably introduce some phase cancellation effects because of the distance between the two mics, but whereas this might be a problem with a natural sound source, it actually becomes another creative possibility with the electric guitar.

TIP: Experiment with changing both the mic position relative to the centre of the speaker, and its distance from the speaker, as these two parameters interact. For example, when close miking you'll hear a very different tonality when the speaker is miked on-axis compared to miking close to its edge, but as you move the mic further away these lateral adjustments will have less influence on the tonality. Despite the number of textbooks telling you to mic the amp very close to the grille, backing off by 200mm or so can often improve the sound.

By changing the distance of the farther mic and adjusting the height of the amp off the floor, you can 'tune' these phase effects until you hear something you like.

Since at least one mic is normally close to the speaker, room acoustics and low-level noise from computers are usually inaudible, so it's perfectly feasible to record and play in the same room while monitoring via speakers rather than headphones. As long as you are close-miking the amplifier and the mic is faced away from the studio monitors, the amount of spill picked up should be very small. When working this way, it's sensible to turn down the DAW monitoring level for the guitar track during recording so that you only hear the guitar from its own amplifier, with just the backing track over the studio monitors. This avoids any latency problems and also allows you to adjust your monitor balance while recording by physically adjusting your position towards the guitar amp for more guitar level or towards the studio monitors for more backing track level.

If you try some of the more distant miking positions, room acoustics and computer fan noise may become an issue. If so, deploy a few duvets or other sound absorbers to tame any reflections that might otherwise reach the microphone. Propping a mattress or folded duvet across a corner and facing the amp towards it works well and will also help reduce the amount of spill from the guitar amp that gets into other mics in the room, if you are recording two or more parts at the same time.

Remote Amping

You can also record with the amplifier in the live room (or even in a cupboard or buried beneath a pile of blankets) while you stay in the control room. This has the benefit that what you hear over the studio monitors is the exact sound being recorded – although you lose out on the acoustic interaction between the guitar and speaker, which can be an issue with high-energy rock sounds or performances that rely on controlled feedback. We've taken this approach on several *Studio SOS* jobs by running cables out to a guitar speaker

A DI'd guitar part can be fed back out to a guitar amplifier (which can then be miked normally and re-recorded), using a 're-amping' box, which takes the balanced, line-level output from the computer interface and provides an instrument-level unbalanced output to feed the guitar amplifier.

placed in a hallway or behind a sofa in a different room in the house with blankets draped over it.

Re-amping

A guitar part recorded 'dry' can of course be processed using an amp modelling plug-in, but an alternative approach is to feed it back out to a guitar amplifier, which can then be miked normally and re-recorded. Re-amping, as this technique is known, is a useful way to change the guitar sound if the one you originally chose didn't work out in the context of the final mix.

Of course, guitar amps are designed for the low-level, unbalanced, high-impedance signal from an electric guitar, not the line-level, balanced signal from a computer interface. The ideal solution is to use a 're-amping' box, which is a bit like a passive DI box in reverse: it takes the balanced line level output from the computer interface, attenuates and unbalances it, and provides an instrument-level output to feed the guitar amplifier. Usually there is also a ground-lift facility to avoid ground loops, and sometimes a level control to help optimise the level matching. If you don't have a re-amping box, you can take an unbalanced line out from your audio interface directly to a guitar amp input, but you'll need to adjust the output level so that the signal going in to the amp is similar to the level you get when feeding it from your guitar.

Hiss and Hum

Hiss is a fact of life for guitar players whenever distortion is used because the high-gain involved boosts any inherent circuit noise and pickup of extraneous interference along with the wanted guitar sound. In a number of cases we've found that the level of background noise is far higher than it needs to be simply because the player is sitting close to a computer. The electrical interference generated by most computer systems can significantly add your guitar/amp noise, so it's always best to stand well away from your computer while recording.

Hum is a particular problem when working with guitars that are fitted with single coil pickups, such as Fender's classic Stratocaster and Telecaster models, as their pickup coils are susceptible to stray magnetic fields from any other electrical equipment nearby.

Putting a noise gate plug-in on the guitar track after recording can help kill any remaining hiss and hum during pauses in the playing, where the noise is most obvious, but it is important that any gating is done before adding reverb or delay, otherwise the action of the gate will tend to mute the tail end of the reverb or echo decay effect. It is also a good plan to set the attenuation amount on the noise gate to pull down the noise by say 10 to 15dB rather than killing it altogether, as the modest attenuation sounds more natural on any exposed guitar parts.

/ NOISE-REDUCING GUITAR MODIFICATIONS

There are after-market humbucking replacement pickups for most guitars that are much less susceptible to hum, although purists will tell you that even the best of these changes the tone of the guitar slightly, so the choice is a personal one.

If you like to indulge in DIY you can often make a useful improvement to guitars fitted with single-coil pickups by lining the body cavities of the guitar with self-adhesive copper foil to form an electrostatic screen. The separate pieces of foil should be overlapped and soldered at one or more points to maintain electrical conductivity and then linked to the ground of the output jack socket or the rear of the pots using a short piece of insulated wire. The foil forms an effective screen around the pickups, wiring and controls, reducing the amount of buzz picked up. You can buy suitable self-adhesive copper foil from most guitar parts specialists and it can be cut to shape easily using scissors. Conductive paint, also available from some guitar parts specialists, is a practical alternative to copper foil.

However, this technique won't help to reduce hum, which is induced magnetically directly into the coils – the only solution here is to use replacement, stacked humbuckers or a bespoke, after-market hum-cancelling solution (see picture).

If you use copper foil for screening, make sure that there is electrical continuity across all the individual pieces, and also that there is a ground wire attached to somewhere!

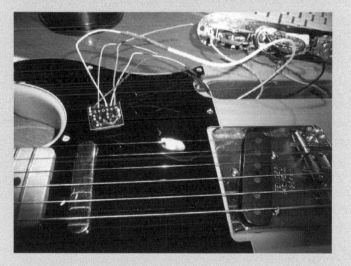

The one really successful, although not cheap, way to reduce noise from a single-coil instrument without changing the sound is to fit an Illitch Electronics coil. This is a special hum-cancelling coil that resides inside the tremolo back-plate in the case of Stratocasters, or the scratchplate, in the case of Telecasters. They are highly effective, but are only available for a limited range of guitar models at present.

Where modification of the guitar isn't viable, the only practical solution is to sit as far away as possible from sources of interference, such as amplifier power transformers and computers, and then to find the orientation within the room at which the guitar picks up the least amount of hum and interference. Some professional engineers get the player to turn until the hum is at a minimum, then they tape a line on the floor so the player always knows exactly which way to point the guitar for the cleanest results. Old fashioned CRT (cathode ray tube) computer displays are also a common source of buzz, so switch these off when recording, or replace with flat LCD screens, which don't cause hum and buzz problems.

Other gear powered by external power supplies or containing large internal power transformers can also cause interference, so switch off anything not being used and try to keep as far away as possible from anything that seems to be giving problems. Lighting dimmers, fluorescent lights and some fan-speed controllers cause an annoying buzz rather than hum, so either turn these off or set them to maximum to alleviate the problem.

Bass Guitar

For many kinds of music, a clean DI'd bass guitar signal with a touch of EQ and compression is all that is needed. However, if you are into rock territory you'll find that the 'dirt' introduced by a bass amp and speaker actually helps the bass to stay audible and keeps it sounding punchy. You can sometimes improve the sound of a DI'd bass by adding a modest degree of overdrive using an amp-modelling plug-in, and most guitar amp plug-ins can also produce useful bass sounds, although a dedicated bass version will always give you a bit more tonal scope.

Equally, you can record the bass part through a dedicated bass preamp or bass amp-modelling processor, or even a six-string-guitar-modelling preamp, as they will often have several suitable amp and speaker models that will give a good bass sound. The main difference between the hardware and plug-in approaches

TIP: We've met players who have screened their guitar with foil, but then failed to ground the foil. This usually results in more noise pickup, rather than less, as the foil then acts as an aerial. It is also good practice to cut out a piece of thin plastic packaging to line the base of the cavity to prevent the foil shorting out the back of the pickup selector switch.

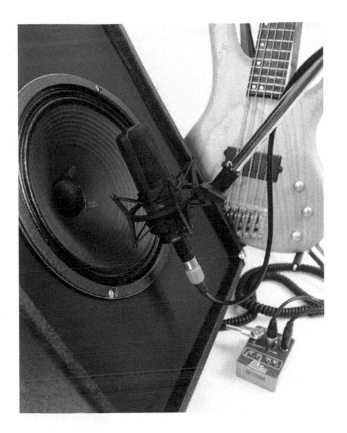

DI'd electric bass is often recorded with just a little EQ and compression, but combining this with an amp signal as well will usually give the sound more character and help it remain audible in a busy mix.

is that computer-based users who have suitable plug-ins get the opportunity to change the sound after recording, whereas if you record through hardware, you are somewhat more limited as to the changes you can make after the event, unless you record a clean DI track at the same time.

If you are miking a bass amp, try starting with a mic position between 6 and 12 inches from the speaker grille and, as with the guitar, move the mic across the speaker cone to find the most suitable tone. Obviously, use a mic with a good bass response, rather one that has a steep low-frequency roll-off with distance from the source (such as is common on vocal mics). But, again, don't be afraid to try whatever you have available as you may be pleasantly surprised. Many large-diaphragm, capacitor vocal mics work well on bass, as long as they don't exhibit excessive low-frequency roll-off.

Dedicated bass-guitar preamps are sometimes used for the character they impart, making a DI recording sound more controlled and complete by blending in a little distortion.

Electrovoice's RE20 and Sennheiser's MD 421 dynamic mics are both popular choices for bass amps, although many large-diaphragm, capacitor models will also work well.

If you prefer the DI option, the main thing to be aware of is that what sounds perfect when heard in isolation may tend to get lost in the mix when the other instruments are faded up. Indeed this is a common problem we've come up against on both our *Studio SOS* and *Mix Rescue* projects. The temptation is to think that a bass sound with lots of powerful low end will sound good in the mix, but in reality it also needs plenty of mid-range too – if only the deep bass was there the bass guitar would be inaudible on most small transistor radios or computer speakers.

Reality Check

Although this chapter has focused on techniques to record electric guitar and bass, a significant number of quality issues we've encountered on our travels have been down to the sound source rather than the method of recording. Guitars that are badly set up and improperly intonated crop up with alarming regularity, so if you are unable to carry out a basic setup yourself, it is well worth having a guitar tech do it for you. Then there are worn and dirty strings, which do the tone no favours and can also affect tuning – and talking of tuning, even some of the better mixes we've been asked to polish have suffered from slightly out-of-tune electric and acoustic guitars, so it's well worth checking your tuning before every take.

With bass guitar, a surprising amount of the final sound comes from the way the instrument is played in the first place. We've had to deal with a lot of bass parts that were played very unevenly, timidly or with excessive fret buzz, and in most cases you can't fix these types of problems with plug-ins or processors. The bass part is so important in most contemporary music that if the bass isn't played well, you're often better off replacing it with suitable bass samples played using a keyboard.

Chapter Ten
Drums

 Most contemporary popular music relies on a rhythm section of some kind, and in pop music this is usually the acoustic drum kit, although some dance genres rely solely on synthesized electronic drum sounds. We've visited a number of small studios that have had problems getting a good acoustic drum sound which is not entirely surprising as most professional drum recordings are made in spaces rather larger than those available to the home studio operator. There are, however, some useful tricks you can try to improve the sound.

Getting a good drum sound in the studio starts with a well-maintained drum kit that's properly tuned and skilfully played. Unfortunately, the latter is not always something you can take for granted – there are countless stories of how a top-name drummer came into a studio, tuned up the house kit, and went on to record some great sounding parts, while later the same day a different drummer playing the same kit without changing anything sounded totally different. How you hit the drums is actually a vital component of the sound, just as much as the precise tuning and damping. If you know any drummers in your area who get a consistently good sound, then ask them to show you how to tune a drum kit as it is an art that many drummers, let alone budding engineers, have yet to fully master.

The other components of a great drum sounds are a musically sympathetic room and appropriate microphones, correctly placed. While the close mics on the individual drums don't pick up much in the way of room sound, the overhead mics are very much influenced by the character of the acoustic space, and that's where many home recordings

Getting a good drum sound in the studio requires a well-maintained drum kit that's properly tuned and skilfully played, plus a musically sympathetic room and appropriate microphones, correctly placed.

run into difficulty. The best professional studios have specially constructed drum rooms, with a high ceiling and well-controlled reflections, but for most of us working in home studios that kind of recording space simply isn't available. The best most of us can hope for is to use acoustic absorbers to remove as much of the room's influence as possible so that you can then use a suitable reverb to add back a more desirable impression of ambience, as a post-production process. This tends to be our typical *Studio SOS* approach whenever we come across a drum kit that is set up in a small room.

As long as the kit to be recorded is well maintained and fitted with good quality heads, it should be possible to get a good sound out of it relatively quickly, so long as the player,

If you are close-miking the drums, you'll almost certainly need to tame excessive ringing with tape or a commercially produced damping product.

or recording engineer, knows how to tune it effectively. Worn drum heads tend to stretch unevenly, causing a loss of tone that tuning and damping can't fix, so always check the condition of the heads before starting any important recording work. Usually, a little careful damping is also needed to tame excessive ringing, but this shouldn't be overdone, as the sound can easily end up being too dead when heard in context. A little Gaffa tape on the head is often all that is needed, though there are also tacky rubber pads such as Moon Gel available from drum stores that do the job more efficiently, with the benefits of being reusable and less messy!

Tuning the Kit

The easiest kit to tune is one that uses single-headed toms, although these are currently not very fashionable.

Double-headed toms are more difficult to tune, but when you get it right they sound wonderful. Of course, you may need to apply damping to both the upper and lower heads to kill excessive ringing but, again, don't overdo it, as what sounds too ringy in isolation may sound perfectly acceptable when the whole kit is playing. If you decide to remove the bottom tom heads for recording you may need to restrain any loose parts, like the nut boxes, which may have a tendency to rattle.

An un-damped tom or kick drum can have an excessively long decay time, and will ring in sympathy whenever other drums are hit, or when bass guitars are played in their vicinity. However, if you apply too much damping the drums end up sounding like old suitcases! Remember that even if the toms seem to ring a little too much every time you hit another drum, this probably won't be a problem when you come to mix as you can either silence the tom signal between hits or insert a gate on each of the tom tracks.

Individual drummers will have their own ideas on tuning but, as a rule, any combination of drum and head has a fairly narrow range over which it will sound its best, and some drum manufacturers now go as far as to print on the drum what note it is intended to be tuned to. If the sound has a dull, 'splatty' quality, with little resonance, the drum head is probably tuned too low in pitch. Conversely, a 'boing-like' resonance usually indicates that the head is tuned too high in pitch.

As a starting point the drum heads should be tensioned as evenly as possible, and you can check this by lightly tapping the head around the edges between the tuners with a drum stick – adjust the tuners until you hear the same pitch all the way around the head. If you need to retune the whole drum to a higher or lower pitch the best way is to adjust opposite tuners in sequence, rather than working your way around the drum from your starting point. Pressing firmly in the centre of a newly-fitted head will help to seat it correctly. A useful technique to create a more dynamic tom sound is to first tune the drum a semitone or so above the pitch you want it to be, then slacken off one tuning lug slightly. This will drop the pitch of the drum again and impart a slight pitch drop after the drum is hit.

Drum tuning is a fairly simple concept, but it's amazing how many drummers, and indeed recording engineers, have no idea how to get the best out of a kit with tuning.

Snares

Most snare drums have 13- or 14-inch metal, synthetic or wooden shells, and they vary in depth with deeper shells tending to produce a deeper tonality. Metal shells give a brighter, more ringing tone, while wooden snare drums tend to sound more mellow and fat. The snare wires must be in good order and properly adjusted to minimise rattling, though eliminating all sympathetic buzzing and rattling is virtually impossible. Some rattles and buzzes are usually accepted as part of the natural sound of the drum kit but, where sympathetic rattles are deemed excessive, the sound can usually be improved by gating the close drum mics. It is usually a mistake to try to damp the snare wires with tape to cure rattling as by the time you've cut down on the sympathetic buzzing, you've probably also killed the tone of the drum.

Kick Drum

Most drummers will have a hole cut in the front head of the kick drum to make miking from the front practical, although today's fashion seems to be for a smaller hole which often makes mic positioning fiddly. In a commercial studio it might be prudent to keep a selection of standard kick drum heads in the popular sizes with larger holes cut into them which could be swapped over prior to recording. However, in a typical home studio it is more likely you'll be using your own kit or working with the same drummer most of the time. In this case you need to ensure that you have a front head hole large enough to allow you to insert the kick drum mic on a short boom arm, and to allow you some flexibility in positioning it. An alternative approach is to remove the front head completely and despite concerns that this can put uneven stress on the drum shell and cause it to distort, I've never known this to happen.

You may need to damp the front head as it can ring on for quite some time, and the standard way of damping a kick drum is to place a folded blanket in the bottom so that the blanket rests up against the edge of the batter head, then you further adjust the position to achieve the desired amount of damping. We've visited more than one home studio where the kick drum has been overstuffed with blankets and cushions or even blocks of foam. You only need enough damping to apply pressure to the bottom quarter of the batter head – any more will kill the natural resonance of the drum causing it to sound thin and lifeless.

Wood, cork, plastic and felt bass-drum beaters all sound different, with the harder materials giving a predictably brighter sound that may be more suitable for some styles of rock music. A commercial pad or even an old plastic credit card taped to the head where the beater hits will further emphasise the beater impact, if this is deemed necessary.

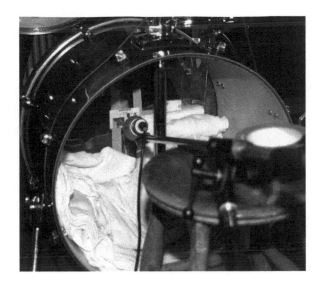

> Your kick drum will almost certainly need some damping inside to achieve a tight, modern drum sound, but too much damping will kill the natural resonance of the drum, causing it to sound thin and lifeless.

The best kick mic position is different for every drum but we often come across kick mics set up too close to the batter head and aimed directly at the point where the beater hits. This tends to produce a gutless 'knocking' sound rather than a satisfying thump, so we suggest starting with the mic halfway between the centre and edge of the shell, initially projecting only 50 to 100mm into the drum. Make a few test recordings and then try moving the mic until you find the 'sweet spot' that combines the right amount of attack and depth.

Specific mics optimised for drum applications are available, but they don't necessarily need to be incredibly expensive or sophisticated. There are some very respectable budget sets available for those on a tight budget, although their kick-drum mics often produce a 'smaller' sound than more upmarket versions. Dynamic mics are often used for the close mics as their inherent slight transient compression sometimes helps to produce a thicker, weightier sound. Other manufacturers favour small electret models as close mics, as these are light and easy to position, particularly in clip-on format. Capacitor mics are usually employed for the overheads as they have to pick up the high-frequencies of the cymbal sound as well as the dynamic attack of the whole kit, although ribbon mics are also popular in many situations.

◄
A good starting point for a kick-drum mic is halfway between the centre and edge of the shell, initially projecting only 50 to 100mm into the drum. Make a few test recordings and then try moving the mic a little until you find the 'sweet spot' that combines the right amount of attack and depth.

The kick mic obviously needs to have a very good low-end response, but most dedicated kick-drum mics actually have a fairly 'lumpy' frequency response to accentuate the 80Hz thump and 3kHz beater-click characteristics associated with the modern pop kick-drum sound. If you use a nominally 'flat' mic, you may need to apply a lot of mid-cut EQ to prevent the sound being 'boxy'. Most engineers seem to choose the AKG D112 or Audix D6 for kick drums, although there are other popular alternatives including the Electrovoice RE20 and Sennheiser MD421. If you try your vocal stage mic on a kick drum you'll still get a sound but you'll probably find you end up with little low-end punch as most dedicated vocal dynamic mics have a low-end roll-off starting as high as 200Hz.

Mic Positioning

There are a number of different ways of miking a drum kit, each capable of producing great results under the right circumstances. If you are lucky enough to have a brilliant-sounding kit in a fabulous room, you may be able to get a great sound with just a single mic (mono or stereo) placed a few feet in front of the kit. However, in the project studio this happens about as often as a squadron of pigs applying for landing clearance, and for any serious work the bare minimum technique involves a pair of overheads, panned slightly left and right to give a suitable stereo image width, and augmented with a kick drum mic.

Where other instruments are playing in the same room, you may need to take some measures to help minimise the spill of unwanted sounds into the drum overhead mics. Solutions might include facing guitar amplifiers away from the drum kit, and/or using power soaks to reduce their volume in the room. Drum overhead mics are typically cardioid-pattern, capacitor models, as these are good at capturing the transient detail of the drum and cymbal sounds, although where spill is not an issue, omni models may give a more open sound. Ribbon mics are also making a comeback in this role, now that more affordable models are available, and being figure-of-eight mics, you can aim their 'dead', 90-degree axis at potential sources of spill, whilst using absorbers behind them. Ribbons usually produce a smoother cymbal sound, too.

Drum overheads are usually set up a metre or so above the kit and spaced around one to one-and-a-half metres apart, although you may sometimes get a better balance by bringing the mics slightly forward of the cymbals to prevent the cymbal sound dominating the drums. You can also use coincident (XY) techniques with cardioid mics for overheads, if preferred. As with any spaced mic technique, care should be taken when positioning the mics to minimise phase cancellation problems.

The above guidelines are all very well and good in a large room, but in a typical home studio the overhead mics will also pick up strong reflections from the walls and ceiling, and these can degrade the sound significantly, making it seem boxy and congested. The most practical option in such cases is to try to remove the natural room sound as much as is practical, using acoustic absorbers such as foam or mineral-wool panels suspended above the overhead mics, and on the walls close to the kit (if necessary). Mattresses propped around the walls can also help. If you can keep the overheads sounding reasonably dry, the illusion of a nice drum room can be added back afterwards using a suitable ambience reverb setting – most DAWs come equipped with a reverb plug-in that will do this job adequately. Some of the software convolution reverbs

TIP: If you set the two overhead microphones at exactly the same distance from the snare drum, you'll ensure that the snare sound doesn't suffer tonal changes caused by phasing issues if the track is played back in mono.

In a good room, with a really good drummer and a well-tuned kit, you can achieve a perfect balance of all the kit elements with just a kick-drum mic and a pair of overheads. Using fewer mics generally results in a cleaner, less coloured sound.

include libraries of real drum rooms, which can sound very flattering without washing the sound out in obvious reverb, but be warned that even the best reverb won't hide a nasty room acoustic that has already been captured in the recorded sound due to lack of treatment of the recording room. So your first priority should be keeping as much reflected sound out of the overheads as possible.

The tactic we have used on many *Studio SOS* visits is to glue foam panels to the ceiling above the overhead mics where this has been permissible, but where gluing is not appropriate we improvise less permanent means of supporting the foam panels such as creating a web of nylon line onto which the foam can be laid, with the nylon cord being attached to hooks screwed into opposite walls. It is also possible to deploy microphone or lighting stands to temporarily hold the absorbing panels just below the ceiling.

Alternatively, ceiling reflections can be eliminated altogether by using boundary mics (sometimes called PZM or Pressure

It's certainly no substitute for proper acoustic treatment, but shielding the rear of the overhead mics from the higher-frequency components in the ceiling reflection, using these easily mounted foam screens, can still offer a worthwhile improvement.

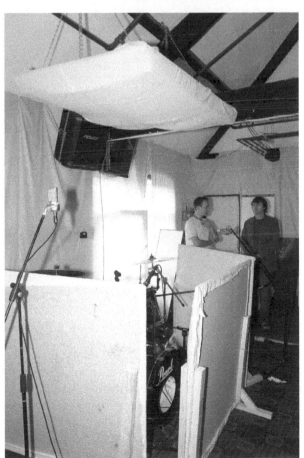

The ideal drum room, as far as many engineers are concerned, will always have a high ceiling, but if you don't have that you can make a big difference to the ceiling reflection by suspending absorption above the kit.

Zone Microphones) fixed to the ceiling above the kit, instead of conventional overheads, so if you record a drum kit on a regular basis and in the same place, this option is well worth considering in rooms with typical domestic-height ceilings.

Using the basic, three-mic technique, you can usually achieve a fairly natural drum sound. However, the obvious compromise is that, other than the kick drum, there is no means of balancing the sounds of the individual elements of the drum kit apart from a little bit of leeway afforded by moving the overhead mic positions slightly. The next step is to add a fourth mic to capture the snare drum directly, since that is the next most important element after the kick. If even more control is needed, then the next step is to close mic all the drums individually – and in this case the overheads may sometimes be high-pass filtered to capture just the cymbals, rather than provide the overall kit sound.

Close miking

For most rock and pop work the basic kick-plus-overheads arrangement is further augmented by close mics on the snare and each of the toms, and sometimes with an additional mic capturing the hi-hat, too. As well as making it easier to balance the kit sound, the close mics are less influenced by the room acoustics and so may make it easier to get an acceptable sound in a less than ideal space.

Some engineers also mic the snare-drum wires from underneath, blending it with the top mic to add more snap and brightness to the sound. Because the under-snare mic is pointing the opposite way to the top head mic, it is essential to reverse the polarity of its output to keep the signals from the top and bottom heads in the same acoustic polarity.

Snare drums and toms can be miked with either dynamic or capacitor mics. A typical starting position for both the snare and tom mics is to arrange the mic 50mm or so above the drum head, just in from the edge and angled towards the centre of the head. Damping pads should be fixed well away from the part of the head the mic is aimed at, and the mic position should be checked to ensure that it isn't in the

Miking both the top and bottom of a snare gives you a lot more options in the mix – blending in a bit more of the under-snare mic is usually a better way of adding more snap and brightness than EQing.

drummer's way when playing. Small changes to the mic position or angle can make a big difference to the sound, so if the tone doesn't seem right at first, change the mic position rather than going straight for the EQ. Some EQ will probably still be necessary when you come to mix your song, but the closer you can get to your target sound by fine-tuning the mic positions, the less EQ you'll need. Mics that clip directly onto the snare and tom rims have become popular in recent years, as these keep the kit free of mic stands and ensure a consistent relationship between the mic and the drum, even if the drum itself shifts position slightly. Just be careful to make sure everything is tight and cables are dressed to avoid creating new rattles.

A separate hi-hat mic is not always necessary, but where required for balance reasons, a small-diaphragm capacitor model, placed around 140 to 200mm from the hi-hat at the side furthest from the rest of the kit, is typical. It is important to ensure the mic doesn't get hit by a blast of air every time the hi-hat closes, so never place the mic looking into the gap between the hi-hats – set it just above or below the level of the cymbals! Where the snare is to be gated to reduce the effect of spill, a separate hi-hat mic may also help avoid the hi-hat level changing as the gate opens and closes, although with a normal balance of overheads and close mics, we've rarely found this to be a problem.

TIP: You can optimise the separation between drums by choosing mic placements that direct the mics away from rather than towards adjacent drums. Similarly, you may wish to angle the snare mic away from the hi-hat to minimise the amount of hi-hat getting onto the snare track.

◄

Sometimes you just want every option covered!

Tracking

Most of the studios we've visited use DAWs rather than hardware recorders, which means there are plenty of tracks available to record all the drums on separate tracks (assuming a suitable multi-channel interface). Where you have fewer recording tracks available you can still get a perfectly acceptable kit sound by mixing the various elements to stereo (or even mono) using a small hardware mixer, and then recording the pre-mixed output. Indeed, this is how many records were made in the past, as anything over four tracks was considered a luxury until the late 1960s – yet some of those records still have great-sounding drums. Having said that, you retain the greatest flexibility if you record each drum and overhead mic to separate tracks, since that gives you more opportunity for gating the individual drums or for manually silencing any ringing between tom hits.

We've demonstrated to many home studio operators that cleaning up tom ringing is actually a very important part of getting a clean, well-focused drum sound. To confirm this for yourself just try soloing a tom track when the toms aren't playing and listen to how much ring and drone there is! Computer users may wish to use their track level automation facility to turn down the tom track levels except for the brief periods when the toms are being played, as this is usually more reliable than using a gate, and still doesn't change the audio file in an irreversible way. Once you get more confident, though, you may choose to do a destructive edit to silence all the spaces between hits – the outcome is much the same.

➤

It's really easy to manually clean up tom tracks in a DAW, muting the areas where there is nothing but spill – it's probably quicker than trying to set up a noise gate to trigger reliably on every hit!

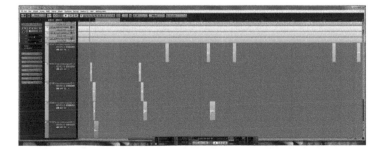

Whether recording individual drums or a pre-mixed track, you should always aim to leave around 12 to 18dB of headroom to keep the transients clean and to avoid running your software mixer levels too hot when all the tracks are combined.

Pre-Mixing and Panning

By convention, the stereo placement of the drums is from the audience's perspective, not from the drummer's, although some producers take the opposite view. The snare and kick are usually panned centrally with the close mics panned to match their instruments' positions in the stereo overheads. A common mistake is to pan some kit elements too far apart, making the drum kit seem unnaturally wide, so use your ears to judge what sounds realistic or appropriate, depending on what else is in the mix.

A suitable 'ambience' reverb on the overheads can put back the sense of a good live drum room, providing you've recorded the tracks as dry as possible by putting up suitable absorbers. A touch of bright plate reverb is often used to liven up the snare, but fashions change so quickly in music that it's worth listening to commercial recordings and focusing on the drum sounds to hear how they've been treated. As you might expect, a common problem we come across when looking at other people's mixes is that they've added too much reverb, which has the effect of filling up all the space, creating a sense of congestion.

You will almost certainly need to EQ the kick and snare to some extent, but if they are well recorded you shouldn't ever need to apply excessive amounts. Furthermore, as long as the kit sounds tonally balanced you can apply more overall EQ, if required, when you hear the drum kit in context with the rest of the tracks. The best way to do this is to route all your drum tracks to a stereo bus, which will allow EQ and compressor plug-ins on the bus to affect the overall kit sound, while the whole kit level can be controlled conveniently using a single fader. If you have a convolution reverb that offers some nice, short, drum-chamber treatments, putting one of these on the drum bus can create a better-integrated drum sound, but it is worth attenuating everything below 200Hz in the reverb send (input) so that the reverb doesn't add too much to the kick drum.

In difficult situations where even your best efforts to control room reflections have been only partially successful, we've managed to arrive at an acceptable sound by applying quite strong low-cut filters to the overhead mics so that you hear mainly the cymbals and the transients of the individual drums being hit – the close mics are used to provide the main body of the sound. A little shelving-EQ boost at around 8kHz injects a bit of life and sparkle into the cymbals. There's more to follow on the subject of mixing drums in the next chapter.

Of course, there are some home studios where there isn't space to mic up a drum kit at all, or perhaps the proximity of neighbours makes bashing around on a drum kit unacceptable. Fortunately, this needn't be a major obstacle to your recording ambitions, as there are other practical options that can produce great results.

Sampled Drums

There are some superb sample-based drum instruments available today, with drum hits recorded at every possible velocity level in world-class studios and using mics that most of us can only dream of owning. But it's not just about the sounds themselves – what you do with them will have a great bearing on the quality of the results you get. Even if you can't play drums yourself, provided that you are able to think like a drummer (no jokes, please!), you can play the samples from a MIDI keyboard in real-time as long as your system is set to a suitably low latency. You probably won't be able to play everything all at once, but it is easy to build the part up in layers. If you can actually play drums recording programmed parts from MIDI drum pads will always result in a more natural feel – full electronic kits and compact pad sets are now available in a range of sizes and prices.

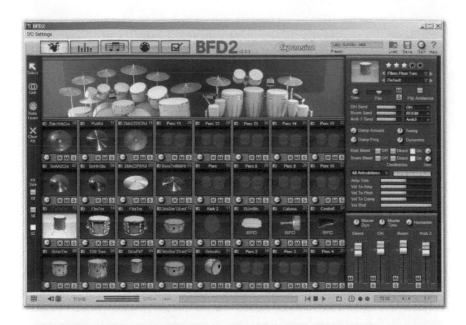

There are some superb sample-based drum instruments available today, including FXPansion's BFD, Toontrack's EZ/
Superior Drummer series.

Divide and Conquer

Of course, there's no reason to use samples for all of your drum parts, or to record them all in the same way. For example, you could program the kick and snare to provide the backbone for the track using samples, and then play in the tom fills and hi-hat patterns afterwards, in separate passes, using real drums and cymbals. If nothing else, this approach eliminates snare rattle, as you don't need to have a snare drum in the room when recording the other parts.

For our own sessions extremely good results have been obtained triggering sampled drum sounds from a set of drum pads, while miking real cymbals and hi-hats at the same time. The sound of the sticks hitting the pads is lost when you put the whole drum sound together and, because sampled cymbals don't respond in the same way, capturing real ride and

▲

The expressive playing surfaces of today's electronic drum kits make them an excellent way to capture high-quality recorded drum parts. They can also be combined with real hi-hats and cymbals to good effect.

hi-hat cymbals reinforce the illusion that the entire kit is real. You need only add a touch of ambience to make the sound of the cymbals match the rest of the kit. We'll often still use sampled crash cymbals, as they are always recorded using high-quality cymbals, with a choice of several different sizes and types, and as they are generally hit only once or twice during a fill, they don't get the opportunity to sound unnaturally repetitive.

Provided that the performance isn't too sloppy, the slight timing variations of a real drum performance can make the whole rhythm track sound more human and interesting, but by using sampled sounds for the main drums you will be able to achieve a far more professional result than you could through miking up an indifferent kit in a small, untreated room.

Drum Replacement

Some sequencing software includes the ability to turn monophonic audio parts into MIDI data and I've found these to be particularly useful for generating MIDI parts from individual drum tracks. Dedicated drum-replacement software is also available that can detect drum hits and trigger samples automatically. Using this technique allows you to record drum parts using a real kit, then, provided that you have enough separation between the drums, you can use the recorded hits to trigger high-quality samples to replace or layer on top of the original recorded sounds.

Most drum replacement software uses a gate-like threshold system to detect hits, with a further setting that prevents a new note being generated for a user-defined time. This 'trigger inhibit time' prevents double-triggering caused by spill or stick bounce. We've used such software many times during our *Mix Rescue* series to replace indifferent drum recordings with suitable-sounding samples, and the beauty of working this way is that the feel and hit intensity of the original performance is retained. The only catch is that the drums generally need to be on separate tracks – although you can sometimes separate out hits individually – and the level of spill must be reasonably low to allow reliable triggering. In most cases, we still retain

> Drum replacement or layering has become a mainstay of contemporary music recording, allowing you to retain the feel and detail of a real drummer's performance, whilst beefing up the key sounds, such as kick and snare, with samples that were recorded under optimum conditions.

a low-cut-filtered version of the overhead track to provide the cymbals as it isn't possible to separate sounds out of the overhead mics to any satisfactory degree. Where the overhead recordings are too poor to rescue, a more pragmatic solution may simply be to replay the cymbal parts from a keyboard using suitable samples.

Using Loops

In addition to sample-based instruments that provide only single hits, there's a wealth of library material using complete drum performances recorded as rhythmic loops of one or two bars. Loops based on recordings of real drum performances always sound very natural with a great feel, and it is usually possible to change the tempo of a loop over a useful range without it sounding too processed. By combining variations on the same rhythm with loops containing drum fills it is possible to create a plausible drum part for an entire song section without any practical knowledge of drumming, and this is certainly a practical option for the home-studio musician who doesn't feel able to devise their own drum parts. Many sampled drum instruments also come with drag-and-drop MIDI files of drum rhythms, which has the advantage that the MIDI data can be edited so that off-the-shelf rhythm patterns can be adapted to fit your own songs exactly, rather than you having to make do with whatever comes closest. There's also no sound-quality penalty for changing the tempo of a MIDI drum pattern.

Summary

Recording drums in a less than optimal space invariably involves compromises, even where the kit is well tuned and well played. Wall and ceiling reflections getting into the overhead mics cause the greatest problems, so this issue should be tackled first. If the results still don't come up to expectations, there's the option of using drum-replacement software to substitute high-quality samples in place of the existing drum hits.

Another valid approach is to use drum pads to trigger drum samples, and to use these instead of a conventional drum kit. These may be used to provide all the drum sounds, although using them in combination with real hi-hats and ride cymbals usually results in a more believable overall sound. Non-drummers will usually achieve better results by using commercial MIDI drum patterns to trigger drum samples, or using pre-recorded audio drum loops in a format that allows some tempo adjustment.

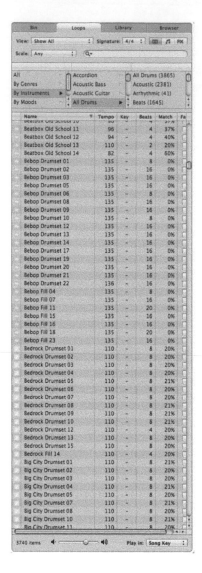

Name	Tempo	Key	Beats	Match	Fa
Beatbox Old School 10	90	-	4	37%	
Beatbox Old School 11	96	-	4	37%	
Beatbox Old School 12	94	-	4	40%	
Beatbox Old School 13	110	-	2	20%	
Beatbox Old School 14	82	-	4	60%	
Bebop Drumset 01	135	-	8	0%	
Bebop Drumset 02	135	-	16	0%	
Bebop Drumset 03	135	-	16	0%	
Bebop Drumset 05	135	-	16	0%	
Bebop Drumset 06	135	-	8	0%	
Bebop Drumset 08	135	-	16	0%	
Bebop Drumset 09	135	-	16	0%	
Bebop Drumset 10	135	-	8	0%	
Bebop Drumset 12	135	-	16	0%	
Bebop Drumset 13	135	-	16	0%	
Bebop Drumset 14	135	-	16	0%	
Bebop Drumset 17	135	-	16	0%	
Bebop Drumset 19	135	-	16	0%	
Bebop Drumset 20	135	-	16	0%	
Bebop Drumset 21	135	-	16	0%	
Bebop Drumset 22	136	-	16	0%	
Bebop Fill 04	135	-	8	0%	
Bebop Fill 07	135	-	16	0%	
Bebop Fill 11	135	-	20	0%	
Bebop Fill 15	135	-	16	0%	
Bebop Fill 16	135	-	16	0%	
Bebop Fill 18	135	-	20	0%	
Bebop Fill 23	135	-	16	0%	
Bedrock Drumset 01	110	-	8	20%	
Bedrock Drumset 02	110	-	8	20%	
Bedrock Drumset 03	110	-	8	20%	
Bedrock Drumset 04	110	-	8	20%	
Bedrock Drumset 05	110	-	8	21%	
Bedrock Drumset 06	110	-	8	20%	
Bedrock Drumset 07	110	-	8	20%	
Bedrock Drumset 08	110	-	8	21%	
Bedrock Drumset 09	110	-	8	21%	
Bedrock Drumset 10	110	-	8	21%	
Bedrock Drumset 12	110	-	4	20%	
Bedrock Drumset 13	110	-	8	20%	
Bedrock Drumset 15	110	-	8	20%	
Bedrock Fill 14	110	-	4	20%	
Big City Drumset 01	110	-	8	21%	
Big City Drumset 02	110	-	8	20%	
Big City Drumset 03	110	-	8	20%	
Big City Drumset 04	110	-	8	21%	
Big City Drumset 05	110	-	8	20%	
Big City Drumset 07	110	-	8	21%	
Big City Drumset 08	110	-	8	20%	
Big City Drumset 10	110	-	8	21%	
Big City Drumset 11	110	-	8	20%	

▲

There's a vast array of complete drum performances recorded as rhythmic loops of one or two bars or eight-bar sections for you to choose from, to use as a guide part to get you started, or as the rhythmic basis of your track.

Chapter Eleven
Mixing

Our *Studio SOS* visits don't just involve improving the acoustic environment of the studio, as we are often also asked to diagnose problems with on-going mix projects. Our popular *Mix Rescue* column goes into forensic detail on a specific track each month, but there are some basic principles that are worth observing when mixing that we can pass on in this chapter, along with many practical tips and techniques.

Recording and mixing can be difficult for musicians without much studio experience, as the way a song is arranged for recording can be very different from the way it might be treated for a live performance. Commercial recording projects are most often set up with a producer directing the engineer, but in the project-studio world most of us have to double as engineer and producer, and quite often as the performer too. To get a good finished product we need to know a bit about arranging and producing, as well as engineering and performing.

Mixing is a skill that requires experience to develop, so if your first attempts are disappointing, you shouldn't start to think that maybe you can't do it – regard them as learning experiences rather than failures. You can learn a great deal about arranging, recording and mixing simply by listening more carefully to your record collection – all the great secrets of music production are there for the taking! As soon as you start to listen analytically, as opposed to just enjoying the music, you'll start to be able to pick out the different elements and be able to see how they're balanced and arranged, what effects are being used, and what the tonal qualities of the individual parts are. You can then apply these ideas to your own mixes.

Mixing a multitrack recording requires a unique blend of artistic and technical skills – it's no surprise that it takes most people some time before they start feeling confident about their abilities.

The key to a good mix is a bit like the key to good cooking – get good ingredients, use the appropriate amounts, and process them correctly. And, as with cooking, too many flavours can confuse and compromise the end product, so question each element of your musical arrangement to ensure that it is there for a specific purpose. The mixing process will be much less challenging if you have managed to keep spill between instruments to a minimum – if you record one instrument at a time, of course, that won't be a problem – but sometimes spill helps to gel everything together nicely too, so it shouldn't be seen as necessarily a bad thing! It also helps to have left plenty of headroom while recording each source so that nothing clips if somebody sang or played louder than they did during the sound-check. Any hum or interference problems really should have been dealt with at source, before recording, rather than hoping you can sort it out later. Any instrument rattles, buzzes, humming

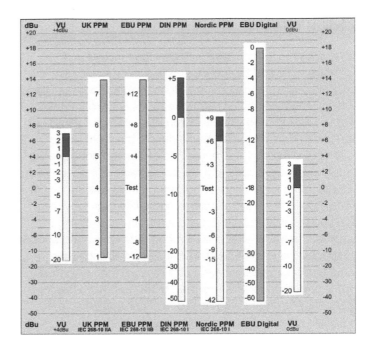

or distortion (other than by intent, of course) that gets onto
your original tracks will be very difficult, and sometimes
impossible, to remove during the mix.

Keep it Clean

When recording audio in a one-room, computer-based studio,
it helps to set up the mics as far away from the computer
as possible and to use acoustic screening (improvised, if
necessary) to further reduce the amount of fan and hard-
drive noise that reaches the mics. Even when the amount of
noise seems quite low, it will become far more obvious if you
subsequently apply compression to a track; every decibel of
gain reduction you introduce knocks off the same amount
from your signal-to-noise ratio. As we explained earlier in the
book, compression also makes room colouration – sometimes
referred to as 'room tone' – more obvious, so minimise this
at source using your improvised absorbers as it is virtually
impossible to correct or to adequately disguise room
resonance in a recorded track after the event.

Getting Started

In the era of tape recording with analogue mixers and hardware outboard the detailed mixing work would only really start when all recording was complete. However, Digital Audio Workstation (DAW) software saves every parameter of your session, and whenever you open the file again it's back exactly as you left it. Some people therefore like to build up the mix as they record, making incremental tweaks throughout the process until the entire track is finished. With this approach there is no actual 'mixing stage' as such, just a continuous process of creation and refinement – by the time you record the last part and apply any processing it requires, the mix is effectively finished. The way a part sounds in a track always depends on what else is playing at the same time, but because each part is heard fully in context as it is added under this scenario, there is perhaps less of a tendency to create parts that don't fit or that have to be radically processed or re-recorded later.

Again, back in the analogue world the studio's 'tape op' would make up a track sheet detailing which instrument was recorded on which track of the tape, along with start and end times. Computer-based DAWs allow you to work without paper (although a notebook is still always valuable), but that doesn't mean you don't need to spend some time labelling things. In addition to naming individual source tracks, you may also

▲

Working with analogue-tape multitracks, it was always essential to keep an accurate log of where everything was recorded using a track sheet (opposite). On a DAW, you can see which tracks have been used and are far less likely to make a costly error, but it still worth naming tracks, channels and buses as you go along, especially if you never want to have to try rebuilding a corrupted session where all the files are labelled 'Untitled Audio'…

wish to keep additional notes using the notepad facility of your computer, and maybe even use a digital camera to capture the control settings on any outboard equipment, such as analogue mixers or processors. If you save these files in the same directory or folder as the audio and project files you'll always know where to find them. In some DAWs you can also assign graphic icons and colours to the tracks to aid navigation around the arrangement page – for example, colour all the vocal tracks yellow, drums brown, all guitars red, keyboards blue, and so on.

Preliminary Housekeeping

Before starting the creative part of the mix – what some people might call 'right-brain' activity – it helps to get as much of the more technical 'left-brain' activity out of the way as possible. A touch of paranoia doesn't go amiss at this stage, either!

As firm believers that digital information doesn't really exist unless copies are stored in at least three physically separate locations, we urge you to back-up any important projects before you start the mix: hard drives can and do fail, usually when it will cause the most grief! This double back-up

▲
Back up your session data, and then back it up again! Hard drives — even good ones like these — can, and do, fail without warning.

approach also enables you to go back to the original files if any destructive waveform edits you perform go wrong. Whether your system is computer-based or a hardware digital recorder you should back-up the files as soon as you can, and update the back-ups regularly as you work through the project. Most computer systems have the ability to allow you to back-up your project using just the specific audio files actually used in the current arrangement, and it makes sense to use this approach to save drive space – there's little point backing up discarded takes unless you think you might change your mind.

If you're working on a collaborative project, ensure where possible that you have the uncompressed WAV or AIFF files to work with. MP3s are fine for sending test mixes back and forth, but you should avoid using MP3 audio files as part of a final mix as their quality has already been compromised by the MP3 data reduction process. This may not sound obvious at the time, but your finished mix will inevitably be converted to an MP3 at some point if you want to make it available as a download, and a double dose of MP3 data reduction can have a very noticeable effect on audio quality.

TIP: Don't rely on backing-up files to a different partition on the same hard drive as you can still lose everything if that drive fails – back-ups must be saved to a physically separate drive. In addition to backing-up to hard-drives, also consider using recordable DVDs, local network storage systems (NAS), or online 'cloud' storage. Apple's Time Machine or one of the Windows automatic backup systems can also save the day when things go badly wrong.

Track 'cleaning'

Before starting the mixing process it's worthwhile going through the individual tracks one at a time, checking for unwanted noises or unintentional audio picked up before the playing starts or after it finishes. Unwanted sounds can be permanently silenced in the waveform editor or turned down using either fader or mute automation. Most of the more sophisticated hardware recorder/workstations include fader and/or mute automation so you can generally do this quite easily. It doesn't matter how you decide to tidy up any unwanted noises as long as the job gets done, although a non-destructive method, such as fader automation, affords the luxury of undoing any mistakes. It is also common practice to mute audio during sections of the track where the instrument or voice isn't playing: for example, muting the vocal track during instrumental bridges or solos, or cleaning up spaces in electric guitar tracks, as these are often quite noisy. With live drums we have already recommended (in Chapter Ten) that you mute the tom tracks between hits, and again it is often more reliable to do this manually using fader automation or destructive editing rather than using gates. It may also be beneficial to gate the kick drum to reduce spill and to cut off any excessive ringing – but we are choosing a gate in this instance as a more pragmatic solution since the kick drum is generally playing throughout the track, whereas the toms are employed only sporadically.

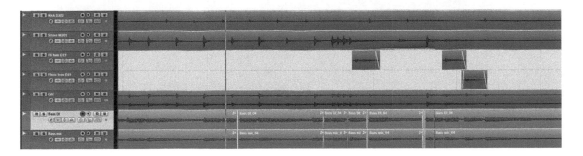

▲

Cleaning up your tracks before you start trying to mix them will allow you to concentrate on the creative process, rather than constantly being distracted by the need for remedial action.

You might also want to check and adjust the audio levels of each individual track to make sure nothing was recorded at too low or too high a level. It is usually easiest to mix a song where the individual tracks peak between −18dBFS and −10dBFS as this leaves adequate headroom for the level to rise as the tracks are combined into your final stereo mix, without any significant system noise (assuming your interface is working with 24-bit conversion).

We see many cases where as soon as you start to build the rhythm-section mix it peaks close to (or even above) the maximum full-scale level, which means the output of the mixer will inevitably clip once other tracks are added (if it isn't clipping already). In theory, the floating-point arithmetic used in typical DAW mixers makes them almost impossible to overload internally, but the output D–A converter will still clip unless you pull down the output fader to reduce the peaks levels at the output below 0dBFS. However, if you have any plug-ins on

If you are recording your source material at 24-bit resolution, there is no advantage to be gained by keeping levels high, in fact you may well be clipping a plug-in somewhere without realising. It makes far more sense to turn up your monitoring, work with plenty of headroom in your software mixer, and make up the final level in the mastering.

your master bus, they may still be clipping, even if the output isn't. All in all, it just makes far more sense to work in the same way as we used to in the analogue domain, with sensible headroom margins. That means moderate source track levels and mix levels well short of digital maximum – and there is no compromise in removing any redundant headroom afterwards in mastering.

Initial Balance

A question we are often asked is how to start balancing a mix. The answer seems to depend on who you talk to! There seem to be two main approaches to getting an initial balance, and the leading engineers and producers we interview seem to be pretty evenly split over which method is best. One method is to start with the rhythm section and then build up the mix one instrument at a time, while the other is to push up all the faders to the unity position and then balance everything in context. Others balance the rhythm section and the vocals (as the two most important elements of the mix) before bringing in the other instruments. In our opinion, for those with less experience, the 'one section at a time' option (starting with drums and bass) might be safest.

Buses and Sends

Having worked on a considerable number of *Studio SOS* and *Mix Rescue* projects where DAW bussing either wasn't used at all, or wasn't used effectively, we thought it might be useful to go over the basics of this relatively straightforward but nevertheless important subject. Routing every source track in your mix directly to the stereo mix-bus makes complex mixes harder to manage, especially if your project contains a large number of source tracks. Effective grouping allows you to control and balance your mix more efficiently using fewer faders.

A related topic, but equally important in mixing, is that of effects buses which are used to allow a single effect plug-in to be shared amongst multiple source tracks. Most DAWs handle

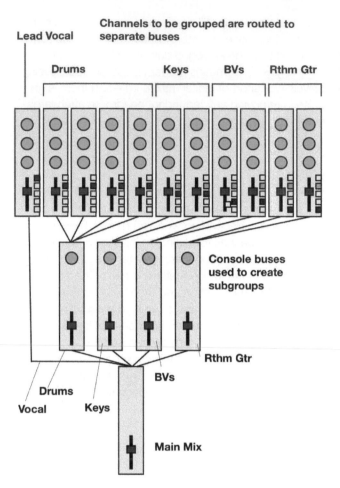

Lead Vocal

Channels to be grouped are routed to separate buses

Drums   **Keys**   **BVs**   **Rthm Gtr**

◄

Routing logical sub-sections of your mix to buses can make a complex mix containing a large number of source tracks much easier to manage, reducing your final balancing process to just a small number of faders.

Console buses used to create subgroups

Rthm Gtr

BVs

Drums

Vocal   **Keys**

Main Mix

bussing in a similar way to their hardware mixer counterparts, so perhaps the place to start is to explain the term 'bus'. In 'mixer speak', a bus is the place to which a number of separate audio signals can be sent to be mixed together – hence the term 'mix-bus'. In a hardware mixer, a mix-bus usually takes the physical form of a length of electrical conductor running the full width of the mixer, with each channel contributing some of its signal to the bus, as required. Just think of it as being like a river (the mix-bus) with streams feeding into it (the mixer channels or DAW tracks), terminating in an estuary where the river flows into the sea, which is analogous to the output that carries the mixed signal.

In a very simple mono mixer there may only be one mix-bus – the single main output mix – and the channel fader determines the channel's contribution to the mix bus. In a basic stereo mixer there will be at least two buses, one for the left output mix and one for the right output mix, with the channel fader setting the overall contribution level and the channel pan-pot deciding how much of the signal goes left and how much goes right.

A larger mixer may have more buses than simply the left and right stereo mix. In the hardware world these additional buses are often called 'sub-groups' and usually feed separate physical outputs so that you can feed different signals or combinations of signals to a recording device or other destination. A popular mixer format might, for example, be 24:8:2, which means the mixer has 24 input channels, eight sub-group buses, and two main outputs (the stereo mix). The individual channels can usually be routed to any desired combination of the eight groups and stereo main outputs, and the groups themselves can also be routed to the stereo mix bus. For example, if you have ten separate channels of drums you can route them all to a sub-group bus (usually a stereo group bus), and then route the output from that mix-bus into the main stereo mix. The drum signals still get to the stereo mix, but as they now go via a sub-group bus the overall level of the drum mix can be controlled using just the bus fader. The advantage of this approach is that once you've balanced the various drum tracks contributing to the sub-group bus, you can then adjust the level of the entire drum kit within the overall song mix using just one fader – and that's a whole lot easier than having to adjust ten faders while trying to keep their relative levels intact whenever you want to turn the drum kit level up or down!

In the software DAW world it is common to have more mix-buses available than we'll ever need, and DAW buses can usually be configured for mono, stereo and even multi-channel applications. DAW mix-bus outputs can usually be routed directly to a physical interface output, or into other mix buses including the main stereo mix. On a hardware mixer this would be done using physical routing buttons, but in a DAW it's usually done via a pull-down menu of possible destinations, and you can think of this almost as setting up a mixer within a mixer.

Using sub-group mix-buses you can set up any number of logical collections of source tracks to allow them to be balanced against other grouped instruments using their respective bus faders. For example, you might put all your drums in one group, all the pad or supporting keyboard parts in another, and all the backing vocals in yet another. Similarly, multiple supporting guitar parts can be sent to their own bus. Main vocals and solo instruments are not normally sent via group buses as these usually require independent control so they can go directly to the main stereo mix bus.

A further advantage of creating sub-groups using bussing is that most DAWs allow you to insert plug-ins into busses, just as you can into channels or into the main mix output. That allows everything sent to one group to be processed together, for example, adding a little overall compression to a multi-layered backing vocal mix to help glue it together, or perhaps some overall EQ to alter the general tonality of the sub-mixed parts.

Auxiliary Sends

Aux sends provide another way of sending signals from a channel to more dedicated mix-buses, and most mixers and DAWs offer the choice of whether the 'send' is derived 'pre-fade' or 'post-fade'. As the names suggest, a pre-fade send picks up its signal before the main channel fader, so its level won't change as you adjust the channel fader. Conversely, a post-fade send is sourced after the channel fader, so as you pull down the channel fader the amount of send signal reduces correspondingly. In a hardware mixer the aux sends usually have a master output level control and appear as a physical output on the connection panel. DAWs are typically more flexible, and these aux or send busses can usually be routed anywhere you like! As with the sub-group buses there is usually a main aux bus level fader so that the overall bus signal level can be changed as necessary.

So, why 'pre-fade' and 'post-fade'? Pre-fade sends, being independent of the channel fader position, are ideal for setting up monitor mixer for the performers. The aux or send bus

Pre-fade sends take their signal from a point before the channel fader and are best used for applications such as monitor feeds – the main mix can then be altered without risk of upsetting the performer(s). Post-fade sends are generally used for effects: with the feed to the effect being taken after the fader, the effect/dry ratio will remain the same, wherever the fader is positioned.

HOW THE MIXER CHANNEL LOOKS

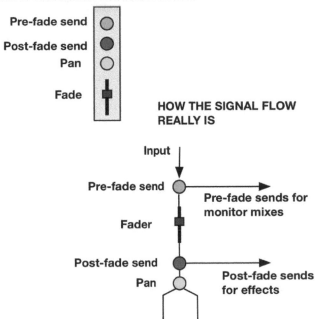

HOW THE SIGNAL FLOW REALLY IS

signal would, in this case, be routed to a physical output on the audio interface (assuming your interface has more than two outputs), and sent from there to a headphone amplifier for the performers to hear the backing tracks. The pre-fade aux send controls in each mixer or DAW channel allows a complete mix to be set up separately from the main stereo mix, allowing different sources to be emphasised or reduced as necessary to help the performer, without upsetting the actual recorded stereo mix. Whereas a small hardware mixer may have only one or two of each kind of send per channel, most DAWs are fairly generous in this area and the number of physical outputs on your audio interface is more likely to set a limit on the number of different monitor or 'foldback' mixes you can set up.

Post-fade sends are usually used to feed effects processors, like reverberation, so that if a source channel is faded out it no longer contributes to the reverb effect. In a hardware mixer the effects units are usually external processors, with the aux send connected to their input(s), and their outputs returned to the stereo mix bus via either dedicated 'effects return' channels, or other spare input channels. When using a DAW, the post-fade sends could be routed to physical outputs on the audio interface to feed external signal processors, but most people use software plug-ins to provide the effects instead. A typical scenario might be to use one of the post-fade sends to feed signals from each channel to a bus into which a reverb plug-in (set to 100% wet, 0% dry) has been inserted, with the bus output (which carries just the reverb signal) being routed into the main stereo mix. The various mainstream DAWs handle this kind of application in slightly different ways, but few stray far from the hardware mixer paradigm. Once set up, this arrangement allows the amount of reverb added to each mixer channel to be controlled using each channel's post-fade send control, and the overall amount of reverb to be set with the bus fader. Obviously each channel will be treated with the same type of reverb – just as each real instrument placed in the same physical recording space will create the same-sounding room reverb – but the amount added to each is fully and independently controllable using the aux sends.

Sends and Groups

A slight complication arises if you set up post-fade sends from channels that are routed through a group bus. This is because the effect level won't change as you adjust the group fader if the output of the effect process is routed directly to the main stereo bus. If you fade down the group, any send effects associated with contributing channels would remain active, and this isn't usually what you want! (Although, you can use this to create the odd special effect where the dry signal fades to zero leaving just a ghostly reverb.) Fortunately, though, this problem is easy to remedy because all you need to do is to route the output from the effect bus back into to the same group so that when you fade down the group, the effects go down with it. The only restriction is that you have to make sure that the effects processor or plug-in is only dealing with signals from channels contributing to the same group bus.

Any effects sent from the source channels routed to a bus will not be attenuated when the bus fader is pulled down, thus inadvertently changing the wet/dry balance — if you have enough computer horsepower, you can solve this potential problem by using a dedicated effect with its returns routed to the same bus.

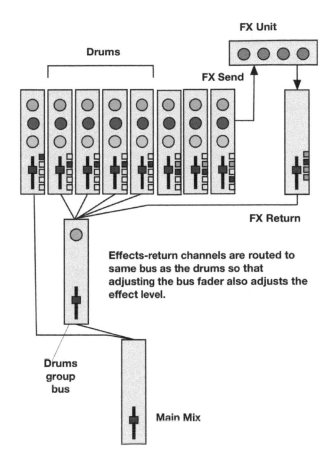

Drums

FX Unit

FX Send

FX Return

Effects-return channels are routed to same bus as the drums so that adjusting the bus fader also adjusts the effect level.

Drums group bus

Main Mix

Where you need the same effect to work on other channels outside of that group you will have to copy the plug-in settings and insert another instance of it in a different effects bus wherever else you need it, but with the horsepower of a modern computer that's rarely a problem.

Grouping the VCA Way

Another way to set up control over specific combinations of channels is to use fader grouping. This is available on some larger analogue and most digital hardware consoles, but is a function that almost all of the mainstream DAWs support. The concept is that instead of routing and physically combining multiple-channel audio signals via a separate mix-bus, the relevant channels are routed directly to the main mix bus, but their faders are linked together to make a 'virtual group' so that they all move together . This means that a number of related channels can be controlled as if they were a single entity, with a single fader, but their audio signals don't have to be combined in an audio sub-group.

Most DAWs work in a way that is roughly equivalent to analogue mixer 'VCA' (or digital mixer 'DCA') groups, where selected channel faders are linked so that they operate together. The process usually involves assigning the required channel faders to a particular fader group – once linked, the group channel faders will all move together whenever any one of them is adjusted, each changing by the same proportional amount to retain the correct relative balance between member tracks. Multiple fader groups can be set up, although most DAWs don't allow the same channel to be a member of more than one group – which is just as well as the result could be quite confusing! With VCA-style fader grouping any post-fade send effects will behave normally, changing in level along with the channel fader, as the 'virtual group' level is now being controlled by collectively changing the individual track faders directly.

The Art of Mixing

During the initial stages of mixing you shouldn't need EQ, panning or effects: you just want to get a feel for how all the parts work together, and to listen for conflicting sounds that may be masking other important elements sharing the same frequency range. The lower midrange between 150Hz and 500Hz is particularly prone to this kind of congestion. Once you've balanced the drums and brought in the bass guitar, you can proceed to bring up the vocals and other instruments, accepting that you might need to adjust the bass or kick sounds later, as they can sound very different when everything else is playing.

If the original parts are all recorded well your rough initial-level balance should sound pretty good as it is, although drums often tend to need a little EQ as a close-miked drum is not really a natural-sounding source at all! Kick drums often need beefing up a bit too, but may also need some low-mid cut to avoid boxiness. It's important that the arrangement leaves adequate space for the vocal, either by leaving temporal space between notes and phrases, or by leaving the vocal part of the frequency spectrum relatively uncluttered – or both. This is largely an arrangement issue, although there are some useful techniques we will explore that can help push the vocals to the front if that turns out to be a problem.

Polishing and Shaving

Once you have a workable initial balance you can scrutinise the various parts to see if anything would benefit from adjustment. If supporting parts, such as keyboard pads or acoustic guitars, are clouding the lower midrange, this can be remedied by thinning them out using low-cut EQ set at a slope of 12 or 18dB/octave. You might, in some instances, be able take the filter cut-off frequency as high as 300 or 400Hz, and although this will leave the instrument sounding very thin when heard in isolation, other parts will be providing the necessary low end in the track and in context it will still sound fine – but you'll now have plenty of spectrum space to work with for the rest of the mix. If you want a part to really sit back in the mix so that it doesn't fight with the vocals and solo instruments you can also take off a little of the higher frequencies, starting around 10kHz and working down until you achieve the desired effect. You'll generally want to use a softer, 6dB/octave slope for this.

Vocal Levelling

Vocals usually sit better in a pop or rock mix if they are compressed to some extent, as compressors not only help level out the differences between the loudest and softest sections, but often also add a musically attractive density to the sound that helps push it to the fore. However, if you try to 'level' a vocal using only compression, you may find that you have to apply far too much compression to the louder sections to keep them under control, and that excessive compression affects the tonal quality as well as emphasising the background noise and room tone. Indeed, too much compression can undo the thickening effect of appropriate compression and allow the sound to slide back into the mix, rather than standing proud in front of it.

So, where the vocal levels fluctuate excessively – as it often does with less experienced vocalists – you'll probably find that you get a better-sounding end result if you first use your track level automation to smooth out the worst variations so that the compressor doesn't have to work quite so hard. However, since fader automation affects the signal after it has passed through any plug-ins in the channel, and we want the level automation to smooth the worst level changes *before* the compressor, one solution is to send the vocal to a bus and insert the compressor there. Alternatively, you can 'bounce' a version of the fader-automation levelled vocal track, and use that one instead in the mix with the compressor now installed back in that (pre-smoothed) channel.

You may also need to automate the overall vocal level throughout the track to help it sit correctly in different parts of the song where the backing level changes – for example,

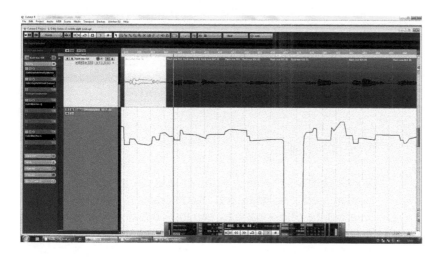

▲

If you want your vocal levels to remain fairly constant, as in most contemporary productions, it is not a good idea to expect the compressor to do all the work – some pre-levelling with fader automation (remembering to insert the compressor after the fader) will always produce a better result.

it may need to be a few decibels louder in the choruses if the arrangement gets busier there, or towards the end of the song where the intensity usually builds up.

When it comes to setting up a compressor for vocals, as a general rule you should adjust the threshold control initially to show about 6dB of gain-reduction, and then listen critically to see if the vocal sounds over-compressed. The action of a compressor depends on the attack and release parameters, the ratio and make-up gain and, most importantly of all, upon the relationship between the threshold control and the level and dynamics of the input signal. There's nothing wrong with using the presets supplied with plug-in compressors so long as you adjust the threshold (or input level) control to achieve the desired amount of gain reduction. If you don't, the plug-in probably won't work in the intended way! This is such an important point to make that we've included a section to cover the use of plug-in presets.

Where EQ is needed to hone the tonal quality of a vocal, try to keep it subtle and avoid using narrow-bandwidth (high-Q) boosts, as these nearly always sound coloured and unnatural. A common, nasal kind of harshness can often be located

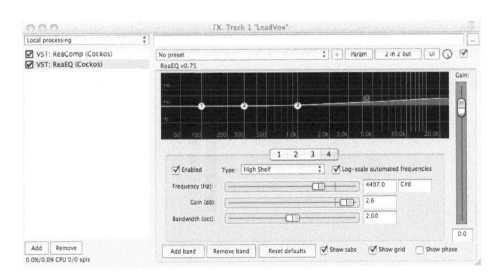

Lifting the whole of the top end with a broad 'air' EQ setting will bring out the detail in a vocal, without making it sound harsh.

around 1kHz, where a little narrow dip can often sweeten things up. A good rule of thumb is to attenuate elements of the sound that you don't like rather than boosting those you feel need to stand out more. If the overall sound is too soft and woolly apply some gentle, shelving low-cut, setting the frequency by ear. Sparkle and breathiness can be enhanced using a little HF shelving boost EQ above 8kHz or so.

/ PLUG-IN PRESETS: GOOD OR BAD?

Presets for effects such as reverb, delay, modulation and pitch changers are often very helpful as you can usually find something that sounds good on a subjective level, and if you tweak the factory settings the results are imme- diately evident. Processing plug-ins, on the other hand, need to be handled with a little more care: as we've discovered on numerous *Studio SOS* visits, problems can and do arise when you start relying on presets for EQ, compres- sion, gating and other 'processor' tasks. For example, EQ presets are created with no knowledge of what your original recorded track sounds like, so the parameter settings are based on assumptions that may be way off the mark. Furthermore, plug-ins relating to dynamic processing, such as compression or gating, have to make assumptions about the average and peak levels of the recorded track – and in many cases they're completely unsuitable without proper adjustment. For example, you may call up a vocal compression preset,

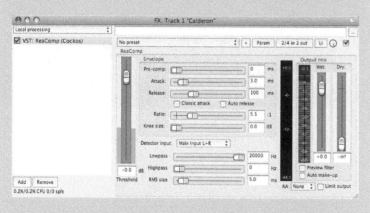

▲

The Threshold setting determines the level at which the compressor will start applying gain reduction – set it too high and the processor won't actually be doing anything.

but if you've left plenty of headroom while recording, as we advised earlier, then the signal level may never get high enough to reach the threshold setting included in that particular preset. We've come across exactly this scenario many times where a compressor has been inserted across a track and yet does absolutely nothing because its threshold is set way too high.

The ratio, attack and release times within the preset are fine for the suggested task but you'll still need to adjust the threshold control to achieve the desired amount of gain reduction. For most tasks, something between 4 and 10dB of gain reduction will do the trick, and there's invariably a gain reduction meter of some kind to show you what affect the threshold control is having on the amount of gain reduction. Once you get the gain-reduction meter showing between 4 and 10dB on the loudest parts of the track, you should be able to fine tune for the optimum amount by ear.

Exactly the same is true of gates and expanders – you'll still need to adjust the threshold control to ensure the gate opens in the presence of signal and shuts

▲

Watch out for software-instrument channel presets that also install a collection of processing plug-ins – not only will this eat up your resources unnecessarily, but also they are extremely unlikely to arrive with the right settings for your mix!

down during pauses without chopping off any of the wanted sound. It wouldn't be so bad if this information was made obvious when you open a dynamics processor plug-in preset, but in most cases inexperienced users are given no indication that further adjustment is necessary.

Also be wary of software-instrument channel presets that also install a collection of processing plug-ins, often ending with a reverb. Not only does this reinforce the impression that it is 'normal' to use lots of different reverb plug-ins within a mix, the EQ added is often designed to make the instrument stand out and sound impressive… and that's not always what's needed.

CASE STUDY – EQ PRESETS

We helped out a reader who wrote and performed some really good songs that were well played and well arranged, but he freely admitted to knowing a lot less about recording and mixing than playing and performing. Because of this he'd used a lot of plug-in presets, including complete instrument and vocal channel presets comprising multiple plug-ins, and had also used a large number of reverb plug-ins in separate channels rather than setting up just one or two different ones on post-fade aux sends.

We listened to some of his mixes and although they seemed pretty well balanced, they were also very harsh-sounding and consequently fatiguing to listen to. There was obviously far too much upper mid-boost – the area we often associate with presence in the 2 to 4kHz range – across the mix, and when we looked at his plug-in settings the cause was obvious: most of his EQ plugins used presets that boosted in exactly this area. Presets are often designed to make the tracks to which they were applied sound big, bright and up-front. In every mix some sounds really do need to be up front, whilst others need to sit behind them in order to create some front-to-back perspective, but the preset designers can't know how your song is arranged. So if you call up lots of presets, you may end up with everything trying to sound big and bright, pushing itself to the front of the mix, and resulting in a harsh flat mix as was the case here. Simply bypassing all the EQ plug-ins produced a marked improvement in the general sound of our case-study mix, and we could then show our reader how to apply the appropriate EQ to the appropriate sources for himself.

EQ presets are often designed for maximum presence and punch, which is fine when that's what you need, but you can't allow everything in your mix to compete for the same sonic space.

TIP: Even when you use plug-in presets just as a starting point, you still need to learn enough about how that particular plug-in works to be able to make further adjustments, according to the needs of the song. Dynamic plug-ins will almost always need you to adjust the threshold setting, while EQ is always best handled on a bespoke basis.

Panning

Stereo panning is an essential element of contemporary music mixing, as many sources are recorded in mono and then panned to suitable positions in the soundstage to create the impression of a stereo sound stage. In a stereo mix panning can also help to improve the aural separation between sounds, but is best not to rely on this as some people still consume a lot of their music in mono. Some mix engineers actually prefer to get their mixes sounding good in mono first (to make sure that any spectral congestion or masking is dealt with from the start), and only then think about the panning – although this usually means the mix levels need to be tweaked very slightly after panning to compensate for the inherent level changes that are imposed by the panning process (the extent being determined by the 'pan law' in use).

The soundstage of a typical pop mix usually approximates the way you might hear the musicians on stage, with the kick drum, snare, bass and lead vocal at the centre and other sources arranged towards the sides. Individual drums are

The panned soundstage of a typical pop mix often approximates the way you might see the players line-up on stage, with the kick drum, snare, bass and lead vocal at the centre and guitars and keyboards arranged further out towards the sides. Toms and stereo overheads will be panned out of centre to give the kit some perspective, but this can sound very unnatural if it is overdone.

usually panned to sound the correct way around from the audience's perspective (although some mix engineers prefer a 'drummer's perspective), and for most musical genres it is important not to pan things too much, making the drum kit unnaturally wide. Close-miked drums should always be panned to match their subjective positions portrayed in the stereo overhead mics to avoid generating confusing and conflicting stereo image information.

Backing vocals and other instruments can be panned wherever your artistic aspirations dictate and, as a rule, you should try to create a fairly even balance left to right – but it's probably wise not to pan any loud, low-frequency instruments too wide. Any instruments producing predominantly low frequencies, in particular the kick drum and bass guitars or synths, are normally panned to the centre so that their energy is shared equally between the two speakers. Stereo reverb effects returns are usually panned hard left and right to create the widest possible sense of space.

Reverb Space

TIP: Adding stereo reverb inevitably 'dilutes' the precise positioning of a sound, so where you want to establish a stronger sense of position, try adding a mono reverb to that sound and then pan the reverb to the same position as the dry sound. This is one occasion where dropping a mono reverb plug-in into a track insert point might be the best option.

Most reverb plug-ins expect a mono input signal but generate a stereo output. In a real acoustic environment the reverberation arrives at the listener from all directions from the side walls, floor and ceiling equally, regardless of where the original signal source is positioned inside the room. So artificial reverb should be the widest element in the mix, and all the other sound sources should be panned in a less extreme way to replicate being located within the acoustic space, in order to sound natural.

For instruments that have been recorded in stereo you need to judge that source's stereo width in the final mix by ear, as much depends on the way the source was miked in the first place. Different stereo mic techniques, and the relative distance between the stereo mic array and the source, will produce different source image widths, but don't worry if you need to turn the pan-pots inwards to reduce the apparent width (or offset them to one side), if that works better in the context of the complete mix.

As a general rule, keep the bass instruments, including kick drums, to the centre of the mix so as to spread the load of these high energy sounds over both speakers.

Don't pan your drum kit too widely. At a typical gig, a physical drum kit probably occupies less than 25% of the width of the stage, so it probably shouldn't appear to occupy much more than this in your final mix.

Lead vocals are traditionally kept in the centre of the mix too, as they are the focus of the performance. You can be more adventurous with panning when you come to the backing vocals.

When you are deciding where to pan an instrument try to conjure up a mental picture of where that instrument should be on an imaginary stage, then close your eyes and adjust the pan control until the sound is where you want it. Go by your ears, not by the numbers around the pan dial.

Don't pan stereo mics or the outputs from stereo electronic instruments too widely or they will appear to take up most of the stage. In particular avoid having a stereo piano where all the bass notes are on one side of the mix and all the treble notes are on the other.

As a rule, pan the outputs from your stereo reverb unit hard left and right and ensure the reverb decays away evenly, without tending to drift towards one side or the other. Mono reverbs can be used to help focus the position of a sound. Delay outputs can be panned wherever your artistic fancy dictates.

/

Mix Perspective

If you rely only on panning for positioning sounds within a mix, the result can sound rather flat with all the sounds sitting along an imaginary line drawn between the two speakers. Achieving a sense of front-to-back perspective is also important in giving your mix interest and scale, but there are no dedicated front/back controls in a typical DAW mixer (unless you're doing a surround mix) and you have to create the illusion of depth using other techniques. We have already pointed out the dangers of trying to optimise each sound in isolation before mixing, and in particular the problems that result from trying to make every track sound as detailed and up-front as possible (especially if you rely on plug-in presets). Everything will be fighting for a place at the front of the mix, leading to a congested and often aggressive sound. The answer is to treat the individual sounds to make the music sound more three-dimensional, with key elements at the front and supporting instruments placed further back.

A fundamental rule of acoustics is that the intensity of the reverberation is essentially similar throughout the room, whereas the level of the direct sound from a source diminishes as you move away from it (according to the 'inverse square law'). That's why sounds heard from further away within a large space seem to be more reverberant. So in a real space the closer sound sources tend to be heard with less reverb and with a greater emphasis on strong early reflections from nearby hard surfaces, while sounds that are further away tend to comprise a larger percentage of reverberant sound, with the emphasis on the diffuse reverb tail.

Also, closer sound sources tend to be brighter while more distant sounds are heard with less high-frequency energy due to air absorption, and you can exploit these effects to enhance the perspective – for example, sounds that you want to appear at the front of the mix can be kept drier and brighter than those set further back. You can go some way towards mimicking these natural characteristics by setting up a couple of aux sends feeding two different reverb settings, one of which is bright and weighted in favour of the early reflections to create a 'forward' sound, and a second that is warmer and more diffuse to create the impression of distance. You can then add these in different proportions to individual tracks to help place them appropriately in the front–back axis. Using an ambience reverb based mainly on strong early reflections helps to reinforce the illusion of closeness, while still adding the necessary 'ear candy' to the vocal sound – and if you listen to the more intimate-sounding records you'll often find that reverb has been used quite sparingly. Conversely, if you want to create a 'stadium rock' effect, where the band is supposed to be a fair distance away from the listener, you can use greater amounts of reverb combined with longer reverb pre-delay times (to make the walls appear further away) to help create this illusion.

EQ to Separate

Even though the 'ideal' is to try to arrange and record all your source sounds so that the mix sounds almost finished just by balancing the channel faders, you can often bring about a further improvement in a mix by the careful use of EQ. As you add more tracks the mix becomes more crowded and may start to sound congested, especially where the inappropriate choice of sounds or a musical arrangement brings several key parts into conflict. Before reaching for the EQ you should listen critically to the arrangement to see what the various parts are contributing, and consider rearranging them (use different chord inversions or orchestrations, for example) or even removing some parts altogether if they are contributing nothing particularly useful to the overall piece. Other parts can be dropped in level to push them further into the background. If that approach doesn't resolve the problem, equalisation gives you the ability to boost or cut the level of some frequency bands so as to change the frequency spectrum of the sound, but you must be aware that an equaliser can only boost frequencies that are actually present in the original sound – EQ is simply frequency-specific gain.

CASE STUDY

We attended one *Studio SOS* session where the musician was using a default sampler sine wave tone for the melody lead line part but was concerned because he wanted a more 'edgy' tonality but couldn't get it to sound any brighter by using EQ. If you take a pure sine wave tone and try to modify it using EQ you'll find that it changes in level but not in tonality – and the reason is that a sine wave has no harmonics or overtones. That means that there's nothing for an equaliser to adjust other than pure volume! EQ only makes sense on harmonically complex sounds where it can change the balance of the various different frequency components that make up that sound. The only practical way of changing the tonality of a sine-wave bass sound is to process it via a distortion plug-in to add new harmonics which will then respond to EQ.

/

If you need to brighten a sound that has no natural high end, you could try a harmonic exciter as these actually synthesize new high-frequency components based on what's happening in the midrange. Provided that you use them in moderation they can actually work very well, and we have found them particularly useful for adding 'bite' to dull-sounding snare drums. Similarly, sub-octave plug-ins can be used to add a lower octave to sounds lacking in deep bass, such as weak kick drums.

EQ 'Bracketing'

EQ can be particularly useful in its role as a mix 'decongestant' when used to narrow down the frequency range occupied by certain instruments. Some engineers call this process 'bracketing', as it uses EQ roll-off at both frequency extremes to 'enclose' the frequency range covered, like brackets around words, and this technique helps reduce the amount by which the spectrum of one sound source overlaps other sounds occupying the same part of the frequency spectrum. For example, many pad sounds are rich in lower midrange frequencies that conflict with the lower end of the male vocal range, while any bright highs may merge with the sound of the guitars or conflict with the vocals. You can often squeeze them into a narrower range, without affecting their role in the track, by using high- and low-pass filters (with 12 or 18dB/octave slopes), or shelving EQ cut at both the high and low ends. Although the EQ'd sound might then seem thin or dull in isolation, you will invariably find that it sits better in the mix. A further benefit

of bracketing is that as the mix becomes less congested, you may then be able to further lower the levels of some supporting sounds without them getting lost.

An alternative technique is to use a parametric EQ to carve a 'dip' in the middle of the spectrum of some sounds to make room for other sounds at a similar frequency. One example of this is using a parametric EQ to place a dip in a bass guitar sound to help keep it separate from the kick drum. You have to find the optimum frequency by ear, but in the case of kick and bass, it's usually in the 100 to 250Hz range.

It is always a good idea to first try to fix a spectrum congestion problem using EQ cut, reducing the level of what you don't want, rather than boosting the bit that you want to hear more, especially if you want to achieve a natural sound. The human hearing mechanism seems to take far less notice of EQ cut than it does of boost, especially when the latter is concentrated in a narrow range. Where you do need to use boost EQ, keeping a wide bandwidth (low Q) sounds more natural than boosting a narrow (high Q) region of the spectrum.

Where the sounds are not natural (such as synthesised sounds or electric guitars), more radical EQ solutions may sound perfectly fine, although sticking to the 'cut first' rule still often produces the best-sounding results. Since these sources have no natural reference, the only rule is that if it sounds right it is right!

Vocals

In a typical pop song the vocals are usually the most important element, so once you have controlled the level using either compression, automation, or both, you need to move on to the matters of EQ, reverb and possibly delay – the main tools in shaping a pop vocal. If you've chosen a suitable mic for your singer you should need very little EQ unless there is something problematic about the voice in the first place. However, here are some strategies that may help you focus your EQ efforts. Even if you've recorded your vocal using a good pop shield some air disturbances or mechanical vibrations may still reach the microphone, so placing a steep low-cut filter at the start of your plug-in processing chain can improve things by removing unwanted sub-sonic rubbish that you probably can't hear, but which will gobble up the headroom and make the speaker cones flap about to no useful effect! If you have a plug-in spectrum analyser it will often reveal unwanted activity in the region below 50Hz. A typical low-cut turnover frequency to employ with a vocal is 80 to 100Hz.

Cutting or boosting in the 2kHz to 5kHz range will allow you to fine-tune the amount of vocal 'presence', although adding significant boost here can cause harshness, too. An alternative strategy that avoids the risk of emphasising potentially aggressive presence frequencies is to apply a broad parametric boost at around 12kHz or an HF shelving boost above 8kHz, often known as 'Air EQ'. Boxiness or any tendency to sound nasal can be improved by applying cuts around 250Hz (boxy) and 1kHz (nasal). You should always try to avoid excessive EQ on a vocal though, as it can all too easily make the voice sound unnatural.

Compression can be applied before or after EQ, but the results will be slightly different depending on the processing order. You get more control if the compressor comes before the equaliser as there is no interaction between the two processes, but putting the EQ first makes the compressor respond more strongly to areas in which you've used EQ boost and less strongly in areas where you've applied EQ cut. Effectively, the compressor is trying to 'level' out any frequency-selective amplitude changes you've imposed with the EQ, although in some situations this is what gives the best subjective sound so it's usually worth trying both options and listening to the difference. However, we'd always advocate putting a low-cut filter before the compressor (or in its side-chain), otherwise there's a risk that it will react to breath blasts and subsonic rumbles rather than to the actual vocal level.

Backing vocals can also benefit from low-end thinning to stop them fighting for attention with the lead vocal and you may also need to add less 'air' EQ so that they sit back a little behind the lead vocal. Other than panning to create the desired stereo image there's no particular special treatment required for backing vocals, although where there are several layers of the same vocal part it can really help to tighten up the sound if you use your computer's editing facilities to line up the timing of some of the phrases (unless of course the original timing

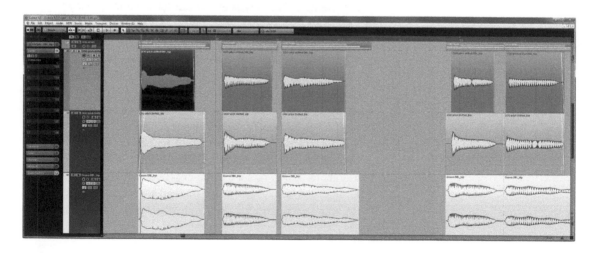

Backing vocals that use a lot of overdubbed parts can be difficult for the singer to keep completely tight. You can make a big difference to their impact in the mix with a little trimming and time alignment of the major consonants.

was spot on), especially where consonants (like 'S' and 'T') are involved. It often helps to make sure that any audible breaths occur at the same place, too.

It can sometimes be helpful to have only one backing vocal part pronounce all of the consonants at the start and end of words during recording. If the others are sung to deliberately de-emphasise these you avoid the untidy effect of three or four 'T's all turning up at slightly different times. You can also fake this effect using level automation to tail off the beginnings and ends of offending words in all but one of the backing-vocal tracks.

Vocal Reverbs

It's often a good idea to set up a specific reverb just for your vocals. Convolution reverbs are brilliant for conjuring up the illusion of a real space, but you'll often find that a good synthetic reverb or plate emulation gives the most flattering vocal effect. Bright reverb sounds are popular on vocals but can also tend to emphasise any sibilance, so you may need to choose a warmer reverb if you detect any problems, or alternatively insert a de-esser plug-in into the reverb send. By de-essing just the reverb input signal, the process is much less obvious than if you de-ess the dry sound. It's also a good idea to roll off the low end, either from the send or the reverb return, to avoid adding more low-mid congestion and clutter to the mix. Many modern vocal effects use a mix of delay (usually with some high-frequency cut to make the repeats less obvious) combined with a suitable reverb, with pre-delay of between 60 and 120ms.

Automatic Tuning

Where the vocal performance is reasonably well pitched an automatic pitch corrector such as Autotune or one of its equivalents can add that final professional polish, and so long as you don't set the correction speed too high there will probably be no audible side-effects. The best results are usually be obtained by setting the correction scale to match the notes being sung, rather than using the default chromatic mode, where it will try to correct every note to the nearest semitone, regardless of whether or not it is in the correct scale. If the song contains a key change you can split the vocal onto different tracks and insert a different instance of the pitch-correction plug-in on each track set up with the correct scale notes for each section.

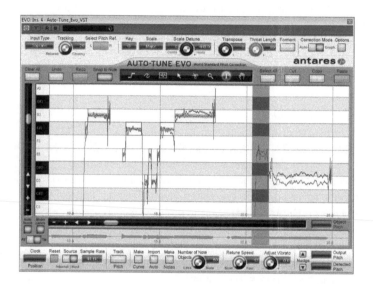

◄

Pitch correction processes like Autotune and Melodyne are fantastically powerful studio tools that can be used for both subtle 'invisible mending' and spectacular creative abuse!

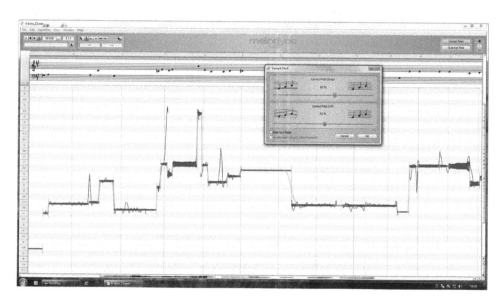

If you 'pitch-correct' a number of double- or multi-tracked vocal parts, the result can sound somewhat 'phasy' as the pitch of each part is now virtually identical. You can avoid this by backing off the pitch-correction speed to different degrees on some of the tracks, or you can apply pitch correction to some layers but not to others. For really serious 'pitch surgery' an off-line pitch-editing solution will provide more precise results.

/ CASE STUDY

We visited one reader who was recording his daughter's singing over some well-arranged and recorded backing tracks. However, we both felt that the use of Autotune was too evident on the vocals and suggested reducing the correction speed. To our surprise he said he hadn't used any pitch correction – that was how his daughter naturally sang! We know now that this is by no means an isolated case – it turns out that many young people sing to emulate what they hear on record and that extends to imitating the processing artefacts of pitch correction!

Bass Guitar

DI'ing a bass guitar may well produce a clean sound with lots of depth, but it can also tend to get lost when the rest of the faders come up. Purpose designed bass-guitar-recording preamps or plug-ins often give better results, as we discussed in Chapter Nine, because they add some of the colouration of a real amplifier and speaker cabinet, giving focus and character to the sound. It is possible to get a good DI'd bass sound by adding compression and EQ, but a dedicated recording preamp or plug-in is often the fastest way to get a sound that works if you don't want to mic a bass speaker cabinet. To some extent, it depends on the type of music you are producing as well, since a DI'd bass may work perfectly well in a sparse arrangement, whereas bass-amp modelling may help the bass to cut through in a busy rock mix.

DI'd bass nearly always benefits from compression to firm it up a bit – an attack time of 20ms or so will help emphasise the transient attack at the start of each note, and the release time can be set anywhere from 50ms to 250ms, depending on the pace of the bass part and on how obvious you want the compression to be. A ratio of about 4:1 and a threshold setting to give a gain reduction of 4 to 10 dB on peaks is a good starting point, but you'll always need to fine tune these parameters by ear, depending on how evenly the bass was played in the first place. On many of the mixes we've encountered in our *Mix Rescue* column, the bass part has suffered from the instrument not being played assertively enough. This tends to result in a sound that lacks punch no matter what processing you use, and fret rattle is also often a problem. Recording can only capture the original performance, so that performance has to be good if the end result is going to be up to standard. It might be better to use a sampled bass rather than a poorly played real bass.

The punch of a bass guitar sound within a mix comes as a combination of its low end and its mid-range. There's very little signal content above about 4kHz other than finger noise, so to increase or decrease the amount of bass, you need to cut or boost between 70 and 120Hz – but remember that boosting the bass end too much will reduce the amount of headroom you have, forcing you to turn the overall bass level down. In fact, a lot of the apparent punch and tonal character of the bass guitar comes in the 200 to 300Hz harmonics range, and the best way to prove this is to listen to your mix on small speakers which have a weak response below 100Hz or so. If the bass seems to vanish from your mix you probably have too much deep bass and not enough going on in the 250Hz harmonics region. Use a low-cut filter to reduce anything below 30Hz or so, as any energy down there will probably be unwanted very low-frequency 'noise' caused by moving the strings slightly above the pickups, just by touching them.

We were asked to improve a mix in which the bass guitar seemed to be getting a little lost. Rather than put it through a bass guitar amp model, we first tried using a simple overdrive plug-in to add just a hint of thickness to the sound, and then applied some EQ boost at around 200Hz to give it more definition in the lower mid-range. The result was surprisingly effective, which shows you don't always need fancy or expensive plug-ins to get the job done.

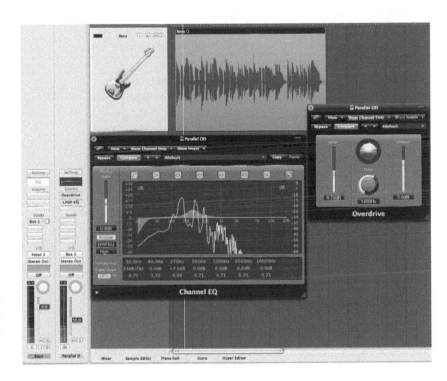

A clean, DI'd bass guitar may sometimes not have enough harmonic complexity to sit comfortably in a mix. A little parallel distortion, with its own EQ, can give it a more interesting midrange without it ever sounding noticeably 'dirty'.

Electric Guitars

A typical electric guitar part covers mainly the 150Hz to 3kHz region of the spectrum – unless you use a clean DI, in which case it covers a lot more of the audible frequency spectrum.

A sharp, low-pass filter cut at 18dB/octave or more can be useful to smooth out a gritty top end without making the instrument sound dull, whilst boosting between 1kHz and 3kHz brings out the natural bite. Be careful not to overdo the latter – things can get very harsh very quickly!

Where there are two electric guitar parts you can use EQ to try to differentiate them, although this is best done at the recording stage with different pickup, chord shape, and amplifier settings. Panning them apart will also help, but a good mix sounds nicely organised in mono as well as in stereo. During our *Mix Rescue* series we've come across a number of tracks where the electric guitar has been so distorted that it has absolutely no dynamics at all. Worse still, if the parts are riffs or chords, played solidly with no breaks, they tend to trample all over everything else producing mid-range sounds, making it very difficult to get a satisfactory mix. You can use EQ to bracket them, which helps, but the real solution is to record using less distortion, and to organise the song arrangement so that there are some spaces and variety. If you need to create a sense of power with more sustain consider using less distortion combined with a compressor.

For ambience, a simulated spring or plate reverb often works best, although if you don't want an obvious reverb effect, try using a hint of convolution reverb with a room, chamber or studio setting. This eliminates the dryness you get in close-miked guitar sounds (real or simulated), and adds a sense of room space but without washing out the sound with obvious reverberation. For 'stadium-style' rock solos, a delay of around 700ms either mixed in with the reverb or used instead of reverb works well.

Modulation effects such as chorus, phasing and flanging can be applied using conventional pedals during the recorded performance, or added afterwards using plug-ins. While plug-ins can usually achieve the right sound guitar players often need to be able to hear an effect whilst playing in order to perform most effectively. If you feel the need to preserve the flexibility to change the sound later, the simplest option is to take a second feed from the guitar as a clean DI and record this to a spare track just in case the effected version isn't quite right. You then have the option to process this clean track in software using an amp-modelling plug-in or re-amp it, to replace or augment the originally recorded guitar track.

Acoustic Guitars

Where the guitar is part of an acoustic-band performance or playing a solo piece you'll probably want to achieve a natural-sounding tonal balance in the recording. A gentle top or bottom cut or boost may be all you need to fine-tune the sound in the mix. If there is any honkiness or boxiness in your recorded sound, however, you can locate it accurately by first setting up a fairly narrow-Q parametric EQ boost, and sweep that across the frequency range of the guitar until the offending part of the sound really jumps out at you. Once located you can apply cut at that frequency to reduce the undesirable element within the sound.

If the recording has been made in a fairly dead room a convolution-based ambience reverb can reintroduce a bit of natural-sounding life to the sound without an audible tail. In a pop mix the low end of acoustic guitars may conflict with other lower mid-range sounds, so it's a good idea to apply a low-cut filter or shelving EQ to thin out the bottom end. This keeps the body sound of the guitar away from the vocals and also stops it blurring into the low end of the electric guitars or the upper reaches of the bass. Listen to an Eagles album to hear how an acoustic guitar can really work well in an electric band context.

Judging the Balance

Level adjustments are almost always necessary during the mix, as the song evolves and builds, even if it is only riding vocal level changes so the compressor has less work to do or lifting guitars slightly during solos, and so on. You can make these small fader movements manually on simpler hardware systems, but mix automation makes this very easy, of course, and on a DAW you can simplify the mix further by putting different sections onto separate tracks. As a rule, avoid changing

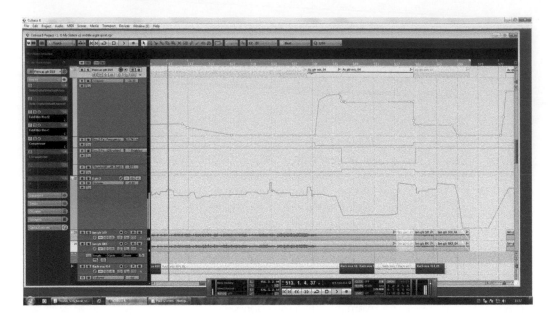

A

Even the best-performed tracks can still benefit from small pushes and fader rides in the mixing, to highlight certain phrases or allow other material to come through in places.

the levels of the drums or bass guitar as these provide the backbone of the track against which level changes in the other parts take place. Changes in intensity in the rhythm section should ideally come from the playing, rather than the mixing.

When you have your basic mix sounding close to how you feel it should be, it is always worth taking a break to listen to a couple of commercial tracks in a similar style. Do bear in mind however that the commercial tracks will also have been mastered, so your mix may sound less tight and punchy at this stage – you can get some idea of how your track might sound when mastered by inserting a compressor and limiter temporarily in the main mix output. Use a low compression ratio of say 1.2:1 and then set the compressor threshold to give you around 4dB of gain reduction so that the track's entire dynamic range is squashed a little bit. Adjust the limiter so that it just catches the peaks giving 1 or 2dB of gain reduction, and if necessary adjust the output level control

If you are comparing your mix to reference tracks, try inserting a temporary mastering chain, consisting of a compressor and a limiter, to give your mix a similar density and loudness to the fully mastered material you are comparing to.

to match the level of your reference tracks. Remember to bypass these plug-ins when you resume mixing, as they are just for comparative listening.

With that frame of reference fresh in your mind you can then make any adjustments to your mix that you feel necessary. Double-check the mix by walking around and listening to it without looking at the computer screen, and do this at several different listening levels, from quiet background music to fairly loud. Also listen to the mix from the next room, or the corridor with the adjoining door left open, as this seems to highlight any serious balance issues remarkably well. A lot of pros do this too, so it is a tip worth remembering. Finally, bounce a rough mix to an MP3 player or CD-R and check how it sounds on the car stereo, portable stereo system and earphones.

Problem Solving

Recording problems can be technical, such as spill, noise and distortion; or musical, such as when two full-on, distorted guitars are both playing busy rhythm parts all the way through a song. And then there are timing and tuning errors, or incorrectly played notes... One of the most prevalent technical problems used to be noise or hiss, although this is far less of an issue now than it was back in the days of analogue tape. Today, hiss is more likely to be due to lack of attention to setting up a proper gain structure, although it can also be generated by older synths, guitar amplifiers, effects pedals and so on. The simplest tool in the fight against hiss is the noise gate, although this has its limitations.

Noise Gate

A noise gate simply attenuates the signal during the pauses between sounds on the track to which it is applied. The amount of attenuation can range from a full mute to a more modest level reduction, depending on the gate's design and control settings. Gating can be very effective, especially where you have a lot of tracks in your mix, so long as the 'wanted' sound is significantly louder than the noise you need to remove. If, for example, a guitar solo appears only once in the middle of a song, you can either trim both ends of the section by editing (the preferred option) or pass the track through a noise gate. The gate is more useful if there are lots of breaks or pauses in the performance, as these will be muted automatically (but would take a long time to edit out manually).

When gating vocal tracks, it is better to set the gate's attenuation (sometimes called range or depth) to between

TIP: Always check your mix on headphones, as these tend to show up noise and distortion problems, as well as any duff edits, more effectively than speakers. They also give you an idea of how the stereo panning will sound on an MP3 player with earphones – one of the main ways music is experienced these days.

You can use a noise gate to automatically clean up your recorded tracks prior to mixing, provided that there is a clear difference between spill and background noise and the level of the material you want to keep. If there isn't, you may be better off using one of the other methods outlined here.

TIP: If you need to use a noise gate in a track that also has a compressor patched in, put the noise gate before the compressor to ensure it triggers reliably. If you put the gate after the compressor there will be less of a differential between the levels of the loud sounds and the quiet ones, so the gate will tend to mis-trigger.

TIP: Reverb or delay can help hide any abrupt note endings caused by gating. If you try to gate after adding reverb or delay, you'll invariably truncate or shorten the decay tail of the effect in an unnatural way – although that is also the basis of the once-popular Phil Collins gated drums effect.

6 and 12dB, rather than allowing it to completely silence the track. This is because we still need to hear the breath noises, but at a reduced level, otherwise the performance can sound very unnatural. Ensure you set suitable hold and release times, so that the ends of the notes aren't cut off abruptly.

Constant broadband noise (hiss or other constant background sounds) can sometimes be dealt with using dedicated de-noising software plug-ins. Some of these work by taking a noise 'fingerprint' of the unwanted sound in isolation, and subtracting that from the wanted signal, whilst others involve a single-stage, multi-band filtering process. There will generally be a control that sets the amount by which the noise is reduced – trying to take it all out invariably introduces odd 'ringing' or 'chirping' artefacts, and better results are usually obtained by aiming for much more modest reduction to the noise of only a few decibels. If the noise is really bad, then doing two or three gentle passes usually sounds much better than one heavy-processing pass. Note that systems relying on a noise 'fingerprint' to calibrate themselves are only effective where the level and frequency

◄

Dedicated digital de-noising processes can be remarkably effective, so long as you don't push them too far, causing side-effects that are more objectionable than the original noise. Several gentle de-noising applications are usually more effective and less noticeable than one heavy-handed application.

spectrum of the noise is reasonably constant. Some of the more sophisticated de-noisers track the noise profile and adapt accordingly, although they can be fooled if the nature of the noise changes too abruptly.

Distortion

While distortion is often used as an effect, we can sometimes face the problem of unintentional distortion. Distortion can either be fairly gentle and musically beneficial, as in analogue overdrive, or very hard and unpleasant as in the case of digital clipping. If not too severe, analogue (harmonic) distortion can be softened by using a very sharp high-cut filter to attenuate the unwanted upper harmonics, but where possible it is better to re-record the part to avoid having to solve the problem in the mix.

DAW users can automate filter plug-ins to apply cut only when the distortion is present, which avoids compromising the clean parts of the recording. Digital clipping is more problematic as it generates anharmonic distortion artefacts that crop up at frequencies both below and above the source frequencies, so they can't be tackled by high-cut filtering. Currently, other than very expensive specialised restoration software, there is no way to properly reconstruct distorted sounds, although it is likely that this technology will eventually percolate down to the project-studio price range where, it could be argued, it is most needed.

Simply the wrong sound

Sometimes you may be presented with a technically good recording, but the sounds still don't seem to work in the context of the mix. Sometimes this is an arrangement issue that needs to be resolved through creative editing, or perhaps discarding or replacing some of the parts, but in other cases you can often use the EQ techniques already described to squeeze the sounds into shape. Where the problem can't be corrected in this way you may find that instruments such as electric guitar, electric bass or drawbar organ can be improved by re-amping. The one problem that EQ never seems to entirely resolve is the over-distorted rhythm guitar part.

/ CASE STUDY

One *Mix Rescue* song we worked on suffered through having a rock organ sound that sounded far too 'polite' and didn't cut through the mix. We added some fairly heavy upper-mid EQ boost and then fed it through an overdrive plug-in which we adjusted to dirty up the sound without making it too fizzy or raspy. This produced a perfectly usable result, but where processing fails re-amping may succeed. Re-amping simply involves routing the offending track to a physical output, then feeding it into a suitable amplifier and speaker that can be miked and re-recorded onto a new track. In this way, the amplifier and speakers modify the tone and allow additional distortion to be introduced where necessary. The same technique can be used to impart a more organic quality to guitar, bass guitar, bass synths, drawbar organs and other keyboard pad sounds.

Although you can feed a line output from the computer interface directly into a typical instrument amplifier (if you take care to reduce the line level signal down to instrument level), a dedicated re-amping box usually gives better results, as it deals with the signal level reduction in a convenient way, sorts out the balanced-to-unbalanced conversion, and provides a ground lift to avoid ground loops. Virtual re-amping is also possible if you use a DAW that has plug-in modelling guitar preamplifiers. By inserting one these on a track, it is easy to reshape a sound in a similar way to passing it through a miked external guitar amplifier and, of course, it is far more convenient.

Adding a small amount of controlled distortion, either from a plug-in or re-amping, can often make a part able to carve out its own space in the mix.

The Mix Medium

Many users record their stereo mix onto new tracks within the DAW, recording the results by 'bouncing' all the tracks to a new stereo audio file, after the mix has been set up and optimised. The use of (16-bit) DAT machines or CD-R recorders as mastering machines is far less prevalent now than it was in the past, as bouncing your mix within the computer is a lot more convenient and allows you to keep your final mix files at a high resolution. If you like to mix on an analogue mixer, either because you like the sound or the more tactile control it offers, the mixer's analogue output can still be recorded back to a new stereo audio track within the computer project. This means there's less requirement for dedicated stereo recorders for this application these days, although there are still plenty of portable stereo digital-audio recorders with 24-bit capability to choose from, if you have a need or preference for mixing onto a hardware platform.

Of course, some people still like to mix to analogue tape, simply for what it does to the sound, and then bounce that mix back into the digital domain as their master. If you are lucky enough to have a good quality two-track analogue tape machine that you can maintain and keep in good working order, then that's certainly a viable option. Alternatively, you could use one of the many tape simulation plug-ins on the market, some of which are excellent, and some rather less so. Beware of any tape-sim that causes audible distortion without being overdriven; that's really not the effect you are after, and not what analogue tape does to the sound at all!

▲

Where an external, hardware stereo recorder is still preferred, a stand-alone solid-state or 'card' recorder capable of 24-bit operation is probably the best solution. The recordings made on such devices can also be backed up to any computer.

Some of the latest tape-emulation plug-ins, such as this one from Slate Digital, make excellent sound polishing and enhancement tools, regardless of whether you think they are accurate or not – and many users will never have actually heard a high-quality analogue tape machine in action!

/ **COMBINING MIX AUTOMATION WITH ADVANCED EDITING**

Not everyone likes to mix with full automation in action across the whole mix, so 'final' mixes are sometimes created by editing together different sections from different mix passes. The desired edit points will usually be on transitions between sections, and it is worth trying to position them shortly before a drum beat so any small discontinuities are likely to be masked by the beat. Cross-fades should be made as short as possible – typically 20ms or so – while still providing a smooth, glitch-free edit, and they shouldn't overlap into drum beats.

Once you've edited a track, listen to it carefully on headphones to ensure that none of the edits are audible, as you can easily miss subtle editing problems when listening on speakers. Listen for and correct any level or tonal changes, as well as more obvious timing irregularities.

A common editing fault we hear is that although the timing of the edit may be correct and the edit made on a suitable beat, the decay of notes played before the edit are missing after the edit point. Once you've created a tight edit, wind back a few bars and then play through your edit, listening to the way the various instruments sound as you pass through the edit point. If there's a problem you may need to choose a different edit point, or perhaps use an asymmetrical cross-fade to achieve the desired effect.

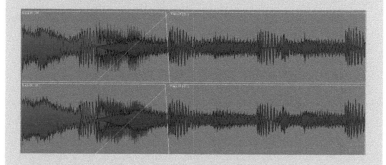

▲

When you have an edit that just won't work, try changing the crossfade parameters.

Chapter Twelve
Mastering

Whenever we visit a home studio, or host a Q&A session somewhere, one question that comes up on a very regular basis is, "what does 'mastering' actually do?" The perception of mastering in music production and delivery has actually changed quite significantly over the last couple of decades, and the job itself has evolved with the way music is distributed and the public's adoption of new ways of listening.

Back when vinyl ruled the earth, the job of the mastering engineer involved equalising tracks recorded at different times to make them sound as though they belonged on the same album, sort out the optimum running order and gaps between tracks, and get the resulting master transferred onto the lacquer. While the first part of the process focuses on the artistic aspects of the album, the latter half is very technical and is about overcoming the limitations of the medium itself. For example, bass sounds had to be moderate in level and in-phase across the two channels of the stereo otherwise the stylus might jump out of the groove, while the amount of top end also had to be controlled carefully taking into account the way the lacquer would 'relax' after cutting, losing HF detail. Also, since louder tracks require wider groove spacing, the knock on is a shorter playing time, so the mastering engineer had to strike a critical balance between both technical and artistic issues to arrive at the optimum solution.

Mastering evolved over the years to develop ways of optimising tracks for radio broadcast, taking into account the effects of FM processing, and as vinyl gave way to CD many of the

The mastering engineer for a vinyl cut has to strike a balance between technical and artistic issues, juggling loudness, playing time and frequency response to arrive at the optimum solution.

previous artistic and technical limitations were removed. New stages were introduced into the process, too, such as the need to encode metadata to define track start points (P and Q coding). Working with digital media also made it possible to make everything sound bigger, louder and brighter than was ever possible with vinyl mastering – and so it was, often to considerable excess!

With the rise of downloads and data-reduced digital formats such as MP3, the mastering engineer's job has again evolved to develop ways of compensating in advance for the effects of 'lossy' data-reduction processes. The prevalence of downloads also now means that a lot of people only put individual hit tracks on their personal MP3 players rather than whole albums, and the ubiquitous 'Shuffle' mode means album tracks are no longer listened to in any specific order, but mastering is still the stage at which an artist's material is organised to form a cohesive album and to share a consistent tonal character, as well as sound great when played in isolation.

Frustratingly, many people still seem to think mastering is mainly about making the tracks sound as loud as possible. In reality, mastering engineers need to take care of all of the technical parameters appropriate to every release medium, as well as apply a final polish and gloss to the sources mixes, correcting for deficiencies at the mixing stage, making the different tracks on an album sound consistent in level and tonality, and setting the correct relative levels and gaps.

Although adding a sense of loudness and power is often a legitimate part of the mastering process, we've come across many project studio mixes (and some commercial ones) that have had all the life squeezed out of them at the mastering stage in an attempt to make them sound louder than the loudest recordings ever released. Not only is an excessive degree of compression and limiting counterproductive – it actually robs the tracks of punch and dynamic interest – it can also make the tracks so fatiguing to listen to that the listener simply reaches for the volume control to turn it down. If the mastered track's waveform looks like a freshly mown lawn, we know we have work to do!

The ubiquity of digital formats led to signal levels being peak-normalised in a way that wasn't possible with analogue media, with hyper-compression being used to squeeze the audio up towards that peak-level ceiling to maximise perceived loudness at the expense of dynamic range. The current fashion for hyper-compressed masters may be about to change, though, with the recent international adoption of loudness-normalisation standards in TV (and shortly radio) broadcasting.

Loudness-normalisation involves the use of new metering systems that measure and quantify the subjective loudness of music rather than just the peak or average levels. Since the absolute levels of different tracks are then matched to give the same subjective loudness regardless of their peak levels, the direct result is that hyper-compression becomes entirely counter-productive and, instead, dynamic range becomes highly desirable once again! Heavily compressed 'loudness at all costs' mixes end up having their level reduced automatically to match the loudness level of other material, and that actually makes them sound considerably weaker than mixes that have been mastered to retain more dynamics.

TIP: The 'Sound Check' feature in Apple's iTunes and iPods works in a very similar way to the international loudness metering standards, automatically adjusting the perceived loudness of tracks to about −16dBFS. There is considerable industry pressure to persuade Apple to turn this facility on by default, and if that happens the 'loudness wars' will end virtually overnight!

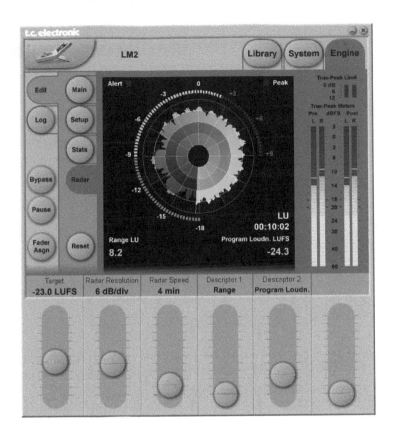

The new loudness-normalisation metering systems measure and quantify the subjective loudness of music, rather than just the peak or average levels, making hyper-compression entirely counter-productive.

Home or Pro Mastering?

It is possible to produce perfectly acceptable mastering results in a home studio provided that you have a reasonably accurate monitoring system, take care over what you are doing, and check your work on as many different speaker systems as possible before signing off on it. While many professional mastering engineers often use exotic analogue equipment to polish the audio, it's possible to achieve just about any processing function you might require using software – either bespoke mastering software or even the plug-ins that come with your standard DAW.

We should say at the outset that if you're planning to have your track or album released commercially, then you are likely to get a better result having it mastered in a reputable commercial

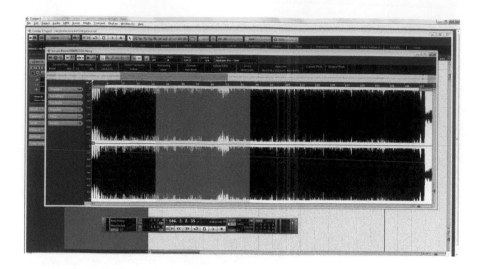

If your mix looks like this before you give it to a mastering engineer, there's almost nothing he'll be able to do to rescue it.

facility. After all, they have the best monitors, set up in rooms that have been designed specifically for mastering – and more importantly, they have a lot of experience. It follows therefore that where professional mastering is envisaged you shouldn't apply your own mastering processes to the finished mix before sending it off. Inappropriate processing is pretty much impossible to undo, and even if you do the job tastefully, the professional mastering engineer almost certainly has better tools and skills to give a more polished-sounding result.

Where a little overall mix bus compression is part of the sound of your mix, though, that should obviously be left in place as the way you balance a song will change depending on the settings of the bus compressor – just remember to mix leaving a little headroom (typically between 3 and 6dB), and don't apply limiting or any EQ to the overall mix.

Evaluation Mastering

Even if you plan to have your material mastered commercially, you may still want to try to set up your own chain of mastering processors just to give you an idea of how your mix might

TIP: It can help the mastering engineer if you send maybe three different mixes with the vocals sitting at slightly different levels in each, as some types of mastering processing can change the subjective level of the vocals. Send one mix that you think is optimised, and at least two more with, for example, the vocals 1dB higher than your original mix and 1dB lower. Some people even send separate 'track minus main vocal' and dry vocal tracks leaving the mastering engineer to balance the vocal as a part of the mastering process (as well as adding suitable compression, EQ and reverb). This feels to us just a bit too much like offloading some of the mix engineer's most important creative decisions onto the mastering engineer, but this approach might work well in some circumstances if the mastering engineer is up for it!

sound alongside commercial records. In this case you'd proceed exactly as if you were mastering the track for real but then switch off the mastering plug-ins before creating the mix that will be sent to the professional mastering house. This is often a useful strategy, as listening to your unmastered mix alongside a commercially mastered record doesn't really give a true comparison.

The Process

There are several quite simple mastering techniques you can employ to make your mixed tracks sound more polished and cohesive, but the most important tool is the ear of the person doing the job. Listen critically to as much music as you can and try to figure out what techniques were used at the mastering stage. Importantly, start out with fresh ears – don't try to master a song straight after spending all day mixing it.

After your own ears, the most important requirement is an accurate monitoring environment, which means very good speakers with a reasonable bass response and a neutral tonality. You also need to work in an acoustically treated space as even the best speakers will produce inaccurate results in a bad-sounding room. A good-quality pair of semi-open headphones will enable you to hear problems that your speakers might miss, and as many people now listen to music on ear buds anyway it is useful to check how your mastered mixes sound when auditioned that way.

We've come across people trying to mix and master on all kinds of inappropriate speakers, from hi-fi speakers with their tweeters poked in, to PA speakers that they also use for gigs! We can't stress too highly the importance of accurate monitoring – without it you really have no idea what your music actually sounds like. If you don't have access to good speakers and a decent room, you're probably better off doing all your mixing and mastering using good quality headphones – but always check your work on speaker systems elsewhere such as friends' studios, the car, the kitchen, the hi-fi...

In a typical home studio you're unlikely to be able to hear any accurate bass below around 60Hz over your monitor speakers, so it is best to avoid applying EQ below that frequency other than to cut unwanted ultra-lows. Headphones can reproduce those low frequencies but in our experience, the subjective level of bass can vary from one listener to another depending on a number of factors, including ear shape and the fit of the headphones. Earbuds vary even more in this respect.

Mastering EQ

Tracks that weren't all mixed at the same time can sound very different, so you may sometimes need to use a little overall EQ on some tracks to make them sit comfortably alongside each other. Mastering EQ is also used to compensate for any fundamental problems in the spectral balance resulting from shortcomings in the monitoring system used for the mixes, such as an excess of deep bass or a recessed midrange.

When trying to get dissimilar mixes to work together, listen first to the bass end of each song to see how that differs and use the EQ to try to get a similar character and weight of bass. The best approach is usually to apply cut in the areas where sounds are too strong before going on to boost where they're too weak, but keep any boosting gentle using a low Q (wide bandwidth) and use cut rather than boost when a more radical change is needed. Also, be prepared to remove any spurious low-frequency information below 30Hz or so using an 18dB/octave low-cut filter if your spectrum analyser shows up significant activity in this area. Muddiness and congestion can often be reduced by applying some EQ cut in the lower midrange, between 150Hz and 400Hz, as appropriate.

If you need to add brightness first try a high-frequency shelving equaliser set to between 8 and 10kHz as this can add a little 'air' to the sound, opening up the mix without making it sound harsh. Alternatively, you can use a parametric EQ set to a wide bandwidth, centred around 10 or 12kHz. If the mix fails to respond to one or two decibels of high-frequency boost it might mean that your original mix is actually lacking high-frequency information – in other words, there might be nothing there for the EQ to lift up!

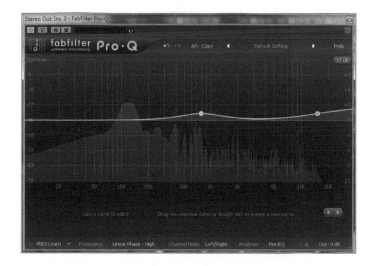

We tend to consider harmonic enhancer/exciters as something of a last resort, but if applied very sparingly in mastering they can enhance the apparent clarity of a difficult mix. However, it's very easy to get carried away as we all like things to be louder and brighter, so as a precaution against over-processing your mix keep a CD player patched into your system and make frequent comparisons between your results and a selection of commercial recordings in a similar style. It is all too easy to lose perspective and start adding unnecessary top boost as your ears get used to the sound.

Whereas headroom is vital when tracking and mixing, very little is needed once a track has been mastered, and so it is common to raise the peak signal level as close as possible to digital full-scale as part of the mastering procedure. Often this is done using a process called normalisation (the file is automatically increased in level so the highest sample peak is at, or close to, digital full scale), but this should only be done after all other processing has been completed, otherwise you risk clipping.

People find it surprising, but even an EQ cut can actually result in an increased level at another frequency, either because of a resonance peak created by the filter slope, or because you may have cut a frequency that was originally partially cancelling something else. In general, we'd advise

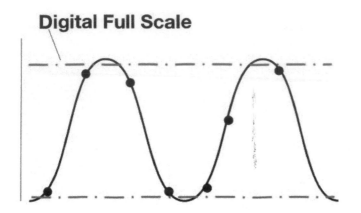

Digital Full Scale

◄

When the audio waveform represented by the samples is reconstructed in the D–A converter, the true peak level can actually end up slightly higher than the sample amplitudes on either side. When reconstructed by the DAC filters, the waveform peaks will be in excess of digital full scale and may suffer clipping within the digital filters.

that you keep sample peaks below –0.3dBFS, and ideally lower still as sample peaks below full scale can still cause overload distortions in some D–A converters and when converting the file to the MP3 format. The reason for this is a phenomena called 'inter-sample peaks' – in essence, when the full audio waveform represented by the samples is reconstructed in the D–A or as part of a complex data-reduction process, the true peak level may well end up slightly higher than the sample amplitudes on either side. In tests, inter-sample peaks have been found as much as 6dB higher than the actual sample values in some circumstances!

The simplest way to adjust the peak level of a track is to place a limiter at the end of the processing chain with its threshold set to around –0.3 to 0.5dB, then adjust the input gain of the limiter so that the appropriate amount of gain reduction is applied to the signal peaks – listening critically and lowering the limiter threshold if you hear any peak distortion.

Mastering Dynamics

Dynamics processors – compressors, limiters, gates, expanders, and similar units – change the level of an audio signal according to its level. To make a track sound louder when it's already peaking close to digital full scale, you might use a compressor followed by a limiter to bring up the average level without raising the peak level. Using a well-designed limiter, you can often increase the average level by 3–4dB, and sometimes more depending on the style of music, without generating any noticeable side-effects. What actually happens is that the top few decibels of short peaks, such as drum transients, are reduced in level which allows the rest of the audio to be increased in volume by the same amount as has been trimmed off the top. Since these peaks are of very short duration and the amount of gain reduction is fairly modest, the limiting is inaudible and the track sounds louder because the average level is now higher.

Importantly, the amount of limiting that can be applied before the process becomes noticeable often depends on how much dynamic processing has been applied to the various elements of the track previously – particularly the drums – when mixing. If the drums are already heavily compressed and limited, the amount of additional limiting you can apply to the entire mix in mastering may be restricted to just 1 or 2dB before the audio quality becomes too compromised. We've come

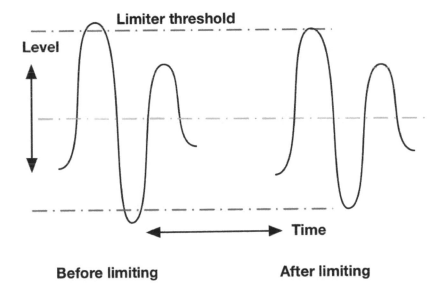

Before limiting   **After limiting**

▲
A limiter reduces the level of the top few decibels of short duration peaks, such as drum transients, which allows the rest of the audio to be increased in volume by the same amount as has been trimmed off the top. After limiting the peak doesn't exceed the limiter threshold, but signals below that level are unaffected. Depending on the limiter design, some distortion may be introduced, as the limiter changes the shape of the signal peaks.

across a scenario in which the initial mix featured compression on individual drums (and often limiting too), with more compression applied via the drum bus and a bus compressor strapped across the whole mix as well. That's three stages of dynamic processing already, so it's little wonder that any further processing in mastering has the capacity to achieve the opposite of what's intended!

Mastering Compression

Using compression on a completed stereo mix can help add energy and even out the differences between the loudest and the quietest parts, making for a more pleasant listening experience in the presence of background noise that might otherwise drown out the quieter passages. Often, though, a compressor will slightly change the apparent balance of a mix, so you may need to use it in combination with a little EQ, placed after the compressor (but before any limiting).

Multi-band compressors were designed to overcome this interaction between the different parts of the audio spectrum that often occurs with conventional, single-band compressors – for example, a high-energy kick drum in a mix can cause very audible gain reduction, taking the rest of the frequency spectrum down every time it hits.

However, most mastering engineers still tend to prefer to use conventional full-band compressors, only resorting to multiband processing to deal with specific problems – for example, a track that becomes aggressively bright only during loud sections can be tamed by setting an upper mid-band compressor to pull down the level of just that part of the spectrum.

Gently Does It

As a rule, very low compressor thresholds used in combination with extremely gentle ratios below 1.5:1 (sometimes as low as 1.1:1) work best for mastering, as this allows an increase in the overall energy of a track and helps to make all the component parts seem better integrated,

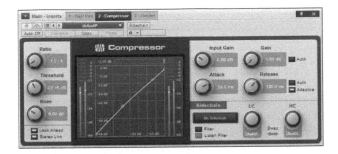

Where compression is used in mastering, it will generally be with a very low ratio, and a very low threshold, so most of the dynamic range will be subject to gentle gain-reduction most of the time.

without making the end result sound obviously processed. With such a low ratio the processing has to start from very low levels, so you'll need to set the threshold level to something like –35dBFS or lower to achieve the optimum 3 to 6dB of overall gain reduction.

This is very different from the normal use of compressors when mixing, where you tend to use a higher ratio and higher threshold, so that only the loudest parts of the signal are affected. The low-ratio mastering approach is far more subtle, as it applies a gentle amount of 'squash' across most of the dynamic range of the material. The practical outcome is that the track level becomes a little more consistent, with the lowest material lifted gently to increase the overall loudness, but the effect of the processing is very transparent and acts more like sonic 'glue' to help the voices and instruments integrate more tightly. If you can hear the compressor working (as opposed to the track just sounding better) you should reduce the ratio or lower the threshold slightly.

It follows from the above descriptions of compression and limiting that if you can apply say 4dB of gain reduction at the compression stage and then another 3dB by limiting the peaks, you'll have brought the average level of the track up by around 7dB. In other words, it should sound considerably louder than simply normalising your mix. However, half a decibel makes negligible difference to the perceived loudness, so setting your limiter threshold to –0.5dB will avoid clipping problems when creating MP3s of your mix without compromising the apparent loudness to any significant degree.

TIP: The dynamics of a song can change between verses, choruses and solos, so choosing a compressor with an automatic release time setting can help keep the processing sound more natural.

Parallel Compression

Many books and articles on compression only describe the use of compressors as in-line processes inserted into channel or mix-bus insert points, rather than as part of an aux send/return loop. In the majority of cases this is exactly what's necessary to get the desired result, of course, but mixing a 'dry', uncompressed signal with a compressed version of itself – referred to as parallel or 'New York' compression – can be a powerful tool, both when mixing and mastering.

To understand the fundamental difference in the results produced by these two configurations, you need to think through what a typical insert point, or 'series-wired' compressor actually does. Compressors have a threshold level below which no processing takes place and above which gain reduction is applied. This means that only the loudest signals are compressed or reduced in level, and signals that don't reach the threshold pass through unaltered.

With parallel compression you would typically choose a high ratio (perhaps 6:1 or above), and a low threshold setting so that the gain reduction meter shows maybe 20 to 25dB of gain reduction during the loudest parts. Listening to the output of this compressor in isolation the signal will sound very obviously over-compressed, but don't be put off! At this stage, you'd normally also adjust the compressor's release time so that any gain pumping, which now will probably be quite noticeable, reinforces the tempo of the track.

The magic happens when this over-compressed signal is mixed with the 'dry', unprocessed signal – transient peaks, such as drum hits, are much louder in the 'dry' channel than in the 'squashed' channel, and hence the dynamics of the drum hits are barely affected, but when the level of the dry signal falls during the softer parts of the track the output of the compressor is still very strong and so makes a significant contribution to the mixed sound.

The overall effect is a mix that sounds more solid and dense, but without the original dynamics of the drum hits being compromised. If this still seems difficult to get your head around, try visualising a range of mountains; a conventional

Compressor

FX Return

FX Send

Main Mix

Parallel compression allows you to raise the average level of your track without the side-effects of attenuating the transient peaks.

compressor reduces the height of the peaks above a certain altitude, whereas a parallel compressor leaves the peaks untouched, but fills in the valleys between them.

/ **CASE STUDY**

We were sent a dance mix to master which was generally very well balanced but lacking some of the punch and depth associated with the genre. After trying conventional compression, EQ and limiting we were still dissatisfied with the end result, so instead decide to give parallel compression a try. In no time at all the track was really pumped up, sounding much bigger and stronger, but losing none of its percussive dynamics. Part of the reason this works is that heavy parallel compression has the effect of 'stretching' bass and kick sounds by keeping up their levels as they decay. Percussive sounds always appear louder if they are made longer, even though their peak level remains the same. /

Editing Tasks

With any tonality and dynamics polishing now taken care of, the next step is to silence any unwanted noise at the beginning and the end of each track, if this hasn't already been taken care of at the mix. You may also want to fade out the very end of the song so that the last decaying sound ends smoothly in silence, but don't start the fade until the natural decay of the sound is well under way or track might appear to end too abruptly.

The next task is to sort out the relative levels between concurrent tracks on an album. This will become less of an issue when loudness normalisation becomes the standard way of working, but while we are still tied to peak-normalisation schemes it's necessary to adjust the relative levels of tracks to create the appropriate listener experience. There's more to matching the relative levels of tracks on an album than giving every song the same subjective level, too – imagine if a heavy rock band track were to be followed by a track featuring a solo singer with an acoustic guitar with both of them at exactly the same perceived level. There are no rules here – this type of decision usually comes down to 'does it feel right'? – but, when you are not sure, a useful guideline is to try to keep the perceived loudness of the vocal consistent between tracks.

There's no point in applying heavy mastering processing to quieter tracks just to achieve more loudness if those tracks are to sit alongside loud songs. All you'd end up doing is reducing the overall level of the quieter songs when you come to build your album playlist. An experienced mastering engineer will anticipate this, and will probably apply less overall compression and limiting to the quieter songs so that their peak levels may still be as high as those of the louder tracks either side, but their subjective level will be lower allowing them to sit more naturally in the running order.

Final Checks

Some DAWs include comprehensive CD-mastering and disc-burning facilities (like Adobe Audition, Sequoia, and SADiE), or you can use a dedicated audio CD-burning program like Roxio's Toast. Our preferred option is to use a CD-burning program that can work with 24-bit source files, reducing the word-length and dithering to 16-bit only after all final level changes and fades have been applied. You'll need to import or arrange the individual 24-bit/44.1kHz, fully-mastered, songs as separate WAV or AIFF files and adjust the gaps and relative levels between tracks as necessary. You are then ready to burn a reference or master production disc.

The CD-burning program must be able to produce a master CD-R disk in 'disc-at-once' mode using the 'Orange Book' format, as this will then allow correctly formatted 'Red Book' discs to be made from it by a disc pressing (replication) plant, or by using CD-R duplication to recordable discs.

Before burning your reference or master CD it is a good idea to listen to the finished playlist all the way through to check that the gaps between tracks feel comfortable. Many CD-burning programs default to a two second track spacing but this can seem too long if the preceding track has a long

TIP: Ensure your individual track files start with at least 160ms of silence before placing them in a playlist – if you trim them exactly to the start of the audio some CD players may miss the very start of the song if that track is selected directly. Also, the very first track on a CD should have at least one second (and ideally two seconds) at the front to give the CD player's laser system a chance to settle before the audio starts. These standard PQ offsets are sometimes taken care of automatically in the CD-mastering program.

TIP: The standard sample rate for mastered audio is 44.1kHz (although there is growing interest in some areas for 'high resolution files' with higher sample rates), so if you are mastering for someone else and have been sent files at a sample rate other than 44.1kHz you should convert them before assembling your playlist. We've had material sent to us with no project information where the original sample rate was actually 48kHz (the standard when working with audio-for-video), and many systems simply play that back around 10% too slowly and a semitone or so lower in pitch!

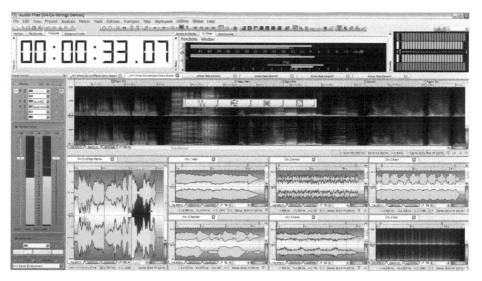

▲

For creating your master CD, you'll need an audio editing program that can work with 24-bit source files, importing the individual fully-mastered songs as separate WAV files, allowing you to adjust the gaps and relative levels between tracks as necessary. This example is Steinberg's powerful Wavelab program.

fade to its final note. This is an artistic choice, though, so just go with what feels right. Once you've burned your first CD, don't make further copies until you've played the master on a few different systems to make sure that both your mixes and the CD itself are OK. In particular, check that each track starts cleanly when selected directly, and any CD-text data is reproduced correctly.

/ **BY THE BOOK**

If the CD is to be used as a master for commercial replication or duplication it will need to be a Red Book-compatible (strictly speaking, Orange Book, as the Red Book standard applies only to the pressed CDs produced from it), PQ-encoded master disk, containing all the necessary track start and pause codes, as well as a table of contents in the form demanded for consumer audio CDs.

A Red Book disc must always be written in 'disk-at-once' mode, rather than 'track-at-once', to avoid data errors being created between tracks. Always use good quality, branded CD-R blanks for important work and use canned, compressed air (available from photographic suppliers) to blow away any dust before burning. Blank CD-Rs are available in different burn-speed ranges, so make sure that the CD-burner speed set in your software is compatible with both the drive and the disc media. It's worth experimenting with different discs and burning speeds to see if one combination proves more reliable than others. Check the writing surface (the underside) of the CD-R is spotlessly clean and free from fingerprints or scratches before attempting to write to it, and always hold the blank CD-R it by its edges, before and after burning.

Cheap disks only save a few pence but may not be reliable – many produce high error rates and may have poor long-term storage characteristics. Some Plextor Plexwriter CD-burners available for PCs include bespoke software that allows the disc error rate to be checked, but we aren't aware of anything similar for Mac platforms.

▲

Use good-quality, branded CD-R blanks for important work and make sure that the CD-burner speed set in your software is compatible with both the drive and the burn-speed range of the disc media.

Always burn two master CD-Rs — ideally using different media — and send both to the pressing plant so that if one is damaged or has an unacceptably high error rate, the other should be usable.

Internet Audio

In addition to making a CD master, you may also wish to prepare versions of your tracks or album for Internet playback or download. The most common audio format for internet

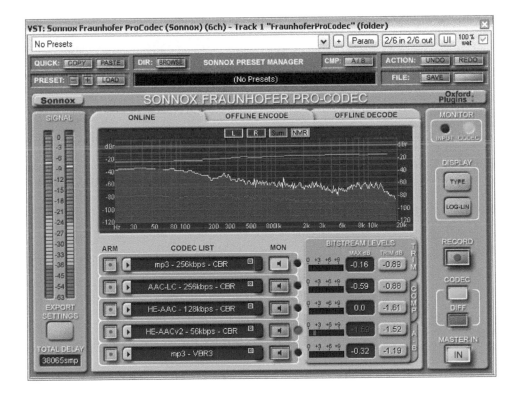

The Fraunhofer Pro-codec plug-in from Sonnox lets you audition the effects of various types and intensities of data compression in real time, which lets you know right away how your carefully mastered audio will sound once set loose in the virtual world.

use is MP3 because its data-reduction system can shrink the size of the audio file to less than a tenth of the original while retaining reasonable quality (at least when using the less extreme settings). MP3 files can be played back universally using computers and hardware MP3 players or smartphones.

In our experience, many musicians still don't appreciate how much the use of heavy data-reduction affects the perceived sonic quality of the music. There are those end-users who will compress everything as much as possible (as this allows more songs to be stored in a given amount of memory), often choosing data rates of less than 96kbps, but in our experience it is best not to compress stereo audio files to less than 192kbps and ideally stick to 256kbps or above if a high subjective quality is to be maintained.

Glossary

Acoustic Foam – A specific type of open-celled expanded polyurethane foam that allows sound waves to enter and flow through the foam, absorbing their energy and preventing them being reflected. The density and depth of the foam affects the frequency range over which it is effective as an absorber.

Acoustic Treatment – A generic term embracing a range of products or constructions intended to absorb, diffuse or reflect sound waves in a controlled manner, with the intention of bestowing a room with an acceptable reverberation time and overall sound character.

Active Loudspeaker or Monitor – A loudspeaker system in which the input signal is passed to a line-level crossover, the suitably filtered outputs of which feed two (or more) power amplifiers, each connected directly to its own drive unit. The line-level crossover and amplifiers are usually (but not always) built in to the loudspeaker cabinet.

Amp/Amplifier – An electrical device that increases the voltage or power of an electrical signal. The amount of amplification can be specified as a multiplication factor (e.g. x10) or in decibels (e.g. 20dB).

Analogue (cf. Digital) – The origin of the term is that the electrical audio signal inside a piece of equipment can be thought of as being 'analogous' to the original acoustic signal. Analogue circuitry uses a continually changing voltage or current to represent the audio signal.

Arming (e.g. for recording) – Arming a track or channel on a recording device places it in a condition where it is ready to record audio when the system is placed in record mode. Unarmed tracks won't record audio even if the system is in record mode. When a track is armed the system monitoring usually auditions the input signal throughout the recording, whereas unarmed tracks usually replay any previously recorded audio.

Audio Interface – A device which acts as the physical bridge between the computer's workstation software and the recording environment. An audio interface usually connects to the computer via FireWire or

USB to pass audio and MIDI data to and from the computer. Audio Interfaces are available with a wide variety of different facilities including microphone preamps, DI inputs, analogue line inputs, ADAT or S/PDIF digital inputs, analogue line and digital outputs, headphone outputs, and so on. The smallest audio interfaces provide just two channels in and out, while the largest may offer 30 or more.

Automation (e.g. of faders) – Automation refers to the ability of a system to store and reproduce a set of control parameters in real time. Fader automation is a system involving moving faders (virtual or physical) in which adjustments made by the user are recorded and can be reproduced in exactly the same way at a later time, or modified if necessary. Most fader, mute, routing and plug-in parameters can be automated in most DAW software.

Auxiliary Sends – A separate output signal derived from an input channel, usually with the option to select a pre or post fader source and to adjust the level. Corresponding auxiliary sends from all channels are bussed together before being made available to feed an internal signal processor or external physical output.

Balanced/Unbalanced Cables – Most audio equipment operates internally with unbalanced signals. These are transferred between devices using a single-core screened cable. The signal voltage is passed on the inner core while a zero volt (ground) reference voltage is conveyed by the outer screen (an all-encompassing metal or conductive plastic braid). The screen also serves to 'catch' any radio frequency interference (RFI) and prevent it from influencing the audio signal.

Where greater protection from electromagnetic interference and freedom from earth references are required, a balanced interface is used. The term 'balanced' refers to identical impedances to ground from each of two signal carrying conductors which are enclosed, again, in an all-embracing overall screen. The screen is grounded, as with the unbalanced interface, but this connection plays no part in passing the audio signal or providing a voltage reference. Instead, the two signal wires provide the reference voltage for each other – the signal is conveyed differentially and the receiver detects the voltage difference between the two wires. Interference instils the same voltage on each wire (common mode) because the impedance to ground is identical for each, and thus any interference is ignored completely.

Signals conveyed over the balanced interface may appear as equal half-level voltages with opposite polarities on each signal wire – the most commonly described technique. However, modern systems are increasingly using a single-sided approach where one wire carries the entire signal voltage and the other a ground reference for it. Some advantages of this technique include less complicated balanced driver stages, and connection to an unbalanced destination still provides the correct signal level, yet the interference rejection properties are unaffected. For interface balancing to provide effective interference rejection, both the sending and receiving devices must have balanced output and input stages respectively.

Back Electret – A form of electrostatic or capacitor microphone. Instead of creating an electrostatic charge within the capacitor capsule with an external DC voltage, an electret microphone employs a special dielectric material which permanently stores a static electric charge. A PTFE film is normally used, and where this is attached to the back plate of the capsule the device is called a 'back electret'. Some very early electret microphones used the dielectric film as the diaphragm but these sounded very poor, which is why later and better designs which used the back electret configuration were specifically denoted as such. Some recent designs attach the PTFE film to the diaphragm and these are known as Front Electrets. Modern electret capsules compare directly in quality with traditional DC-biased capacitor capsules and are available in the same range of configurations – large, medium and small diaphragm sizes, single and dual membrane, fixed or multi-pattern, and so on.

Bass Response – The frequency response of a loudspeaker system at the lower end of the spectrum. The physical size and design of a loudspeaker cabinet and the bass driver (woofer) determine the low frequency extension (the lowest frequency the speaker can reproduce at normal level) and the how quickly the signal level falls below that frequency.

Bass Tip-up – see **Proximity Effect**

Bass trap – A special type of acoustic absorber which is optimised to absorb low frequency sound waves.

Bit Rate (see also Sample Rate) – The number of data bits replayed or transferred in a given period of time (normally one second). Normally expressed in terms of kb/s (kilo bits per second) or Mb/s (mega bits per second). For example, the bit rate of a standard CD is (2 channels x 16 bits per sample x 44.1 thousand samples per second) = 1411.2 kilobits/second. Popular MP3 file format bits rates range from 128kb/s to 320kb/s, while the Dolby Digital 5.1 surround soundtrack on a DVD-Video typically range between 384 and 448kb/s.

Blumlein Array – A stereo coincident microphone technique devices by Alan Blumlein in the early 1930s, employing a pair of microphones with figure-eight polar patterns, mounted at 90 degrees to each other with the two diaphragms vertically aligned.

Boom – A mechanical means of supporting a microphone above a sound source. Many microphone stands are supplied with a 'boom arm' affixed to the top of the stand's main vertical mast. The term may also be applied to larger, remotely controlled microphone supports used in film and TV studios, or even to the handheld 'fishpoles' used by film and TV sound recordists.

Boundary – A physical obstruction to sound waves, such as a wall, or a large solid object. When sound waves reach a boundary they create a high pressure area at the surface.

Buffer (in reference to computer memory and processing) – A buffer is essentially a short-term data storage facility used to accommodate variable data read or write periods, temporarily storing data in sequence until it can be processed or transferred by or to some other part of the system.

Cabinet – The physical construction which encloses and supports the loudspeaker drive units. Usually built of wood or wood composites (although other materials are often used including metal alloys and mineral composites). Cabinets can be 'sealed' or 'vented' in various ways, the precise design influencing the bass and time-domain characteristics.

Cabinet Resonance – Any box-like construction will resonate at one or more frequencies. In the case of a loudspeaker, such resonances are likely to be undesirable as they may obscure or interfere with the wanted sound from the drive units. Cabinets are usually braced and damped internally to minimise resonances.

Capacitor – A passive, two-terminal electrical component which stores energy in the form of an electrostatic field. The terminals are attached to conductive 'plates' which are separated by a non-conductive dielectric. Capacitance is measured in Farads. If a voltage is applied across the terminals of a capacitor a static electric field energy develops across the dielectric, with positive charge collecting on one plate and negative charge on the other. Where the applied voltage is an alternating signal, a capacitor can be thought of as a form of resistance that reduces with increasing signal frequency.

Capsule – An alternative term for a transducer which converts acoustic sound waves into an electrical signal.

Channel – A portion of an audio system dedicated to accommodating a single audio signal. Normally used in the context of an audio mixer, where each channel provides a range of facilities to process a single audio signal (gain, EQ, aux sends, fader etc). A mixer might incorporate 6, 12, 32 or more channels.

Click Track – A rhythmic audio signal, normally comprising clicks or plops, intended as an audible cue to assist musicians in keeping accurate time during a performance. It would not normally be heard by the audience.

Clipping – When an audio signal is allowed to overload the system conveying it, clipping is said to have occurred and severe distortion results. The 'clipping point' is reached when the audio system can no longer accommodate the signal amplitude – either because an analogue signal voltage nears or exceeds the circuitry's power supply voltage, or because a digital sample amplitude exceeds the quantiser's number range. In both cases, the result is that the signal peaks are 'clipped' because the system can't support the peak excursions – a sinewave source signal becomes more like a squarewave. In an analogue system clipping produces strong harmonic distortion artefacts at frequencies above the fundamental. In a digital system those high frequency harmonics cause aliasing which results in anharmonic distortion where the distortion artefacts reproduce at frequencies below the source fundamental. This is why digital clipping sounds so unlike analogue clipping, and is far more unpleasant and less musical.

Clocking – The process of controlling the sample rate of one digital device with an external clock signal derived from another device. In a conventional digital system there must be only one master clock device, with everything else 'clocked' or 'slaved' from that master.

Coincident – A means of arranging two or more directional microphone capsules such that they receive sound waves from all directions at exactly the same time. The varying sensitivity to sound arriving from different directions due to the directional polar patterns means that information about the directions of sound sources is captured in the form of level differences between the capsule outputs. Specific forms of coincident microphones include 'XY' and 'MS' configurations, as well as B-format and Ambisonic arrays. Coincident arrays are entirely mono-compatible because there are no timing differences between channels.

Colouration – A distortion of the natural timbre or frequency response of sound, usually but not always unwanted.

Common Mode Signal – A signal that appears equal in amplitude and polarity on both wires of a balanced interface, and consequently is rejected by the differential receiver.

Comping – Short for 'compilation.' The process of recording the same performance (e.g. a lead vocal) several times on multiple tracks to allow the subsequent selection of the best sections and assembling them to create a 'compilation' performance which would be constructed on a final track.

Compressor – A device (analogue or digital) which is designed to reduce the overall dynamic range of a complex varying audio signals by detecting when that signal exceeds a defined threshold level, and then reducing the amplitude of that portion of signal according to a defined ratio. The speed of response and recovery can usually also be controlled.

Cone – A specific shape of drive unit diaphragm intended to push and pull the air to create acoustic sound waves. Most bass drivers use cone-shaped diaphragms, where the electromagnetic motor of the drive unit is connected to the point of the cone, and its outer diameter is supported by some form of flexible membrane.

Converter – A device which transcodes audio signals between the analogue and digital domains. An analogue-to-digital (A–D) converter accepts an analogue signal and converts it to a digital format, while a digital-to-analogue (D–A) converter does the reverse. The sample rate and wordlength of the digital format is often adjustable, as is the relative amplitude of analogue signal for a given digital level.

CPU – Central Processing Unit – the number-crunching heart of a computer or other data processor.

Crossover – A set of audio filters designed to restrict and control the range of input signal frequencies which are passed to each loudspeaker drive unit. A typical two-way speaker will employ three filters: a high-pass filter allowing only the higher frequencies to feed the tweeter, a low-pass filter that allows only the lower frequencies to feed the woofer, and a second high-pass filter that prevents subsonic signals from damaging the woofer.

Crossover Frequency – The frequency at which one driver ceases to produce most of the sound and a second driver takes over. In the case of a two-way speaker the crossover frequency is usually between 1 and 3kHz.

Daisy Chain – An arrangement of sharing a common data signal between multiple devices. A 'daisy chain' is created by connecting the appropriate output (or through) port of one device to the input of the next. This configuration is often used for connecting multiple MIDI instruments together: the MIDI Out of the master device is connected to the MIDI In of the first slave, then the MIDI Thru of the first slave is connected to the MIDI In of the second slave, and so on... A similar arrangement is often used to share a master word clock sample synchronising signal between digital devices.

DAW – Digital Audio Workstation. Elaborate software running on a bespoke or generic computer platform which is designed to replicate the processes involved in recording, replaying, mixing and processing real or virtual audio signals. Many modern DAWs incorporate MIDI sequencing facilities as well as audio manipulation, a range of effects and sound generation.

dB (deciBel) – The decibel is a method of expressing the ratio between two quantities in a logarithmic fashion. Used when describing audio signal amplitudes because the logarithmic nature matches the logarithmic character of the human sense of hearing. The dB is used when comparing one signal level against another (such as the input and output levels of an amplifier or filter). When the two signal amplitudes are the same, the decibel value is 0dB. If one signal has twice the amplitude of the other the decibel value is +6dB, and if half the size it is −6dB.

When one signal is being compared to a standard reference level the term is supplemented with a suffix letter representing the specific reference. For example, 0dBu means the signal is the same as the standard line-level reference of 0.775mV rms.

The two most common standard level references are +4dBu (1.223V rms) and −10dBV (0.316V rms). The actual level difference between them is close to 12dB.

Decca Tree – A form of 'spaced microphone' arrangement in which three microphone capsules (usually, but not always, with omnidirectional polar patterns) are placed in a large triangular array roughly two metres wide, with the central microphone one metre further forward. Sounds approaching from different directions arrive at each capsule at different times and with slightly different levels, and these timing and level differences are used to convey the directional information in the recording. The timing differences between channels can result in unwanted colouration if they are combined to produce a mono mix.

Decoupler (also isolator) – A device intended to prevent the transmission of physical vibration over a specific frequency range, such as a rubber or foam block.

Delay – The time between a sound or control signal being generated and it auditioned or taking effect, measured in seconds. Often referred to as latency in the context of computer audio interfaces.

DI Box – Direct Injection Box. A device which accepts the signal input from a guitar, bass, or keyboard and conditions it to conform to the requirements of a microphone signal at the output. The output is balanced and with a low source impedance, capable of driving long mic cables. There is usually a facility to break the ground continuity between mic cable and source to avoid unwanted ground-loop noises. Both active and passive versions are available, the former requiring power from internal batteries or phantom power via the mic cable. Active DI boxes generally have higher input impedances than passive types and are generally considered to sound better.

Diaphragm – the movable membrane in a microphone capsule which responds mechanically to variations in the pressure or pressure gradient of sound waves. The mechanical diaphragm vibrations are converted into an electrical signal usually through electromagnetic or electrostatic techniques such as ribbon, moving coil, capacitor or electret devices.

Differential Receiver – A differential receiver is used to convert a balanced input signal to an unbalanced on at the output. It has two inputs which are summed together in opposite polarities so that common mode signals cancel each other out and are ignored, while differential signals are passed intact. A transformer can be used as a passive differential receiver, but most systems use active electronics.

Differential Signal – A signal which is applied between the two wires of a balanced interface, and which consequently is passed by the differential receiver. A differential signal can be presented as two signals of identical amplitude (half the total amplitude each) but opposite polarities – as would appear at the output of a dynamic microphone or a balancing transformer. However, it can also be presented as a full level signal on one line and no signal at all on the other, which is how 'impedance balanced' outputs appear.

Digital (cf. Analogue) – Digital audio circuitry uses discrete voltages or currents to represent the audio signal at specific moments in time (samples). A properly engineered digital system has infinite resolution, the same as an analogue system, but the audio bandwidth is restricted by the sample rate and the signal-noise ratio (or dynamic range) is restricted by the wordlength.

Double-lapped Screen – (Also known as a Reussen screen) The signal-carrying wires in a microphone cable are protected from external electrostatic and RF interference by a 'screen' which is a surrounding conductor connected to earth or ground. The Reussen screen is a specific form of cable screen, comprising two overlapping and counter-wound layers which are unlikely to 'open up' if the cable is bent, yet remain highly flexible

Dynamic – An alternative name for microphones that employ a electromagnetic system for converting diaphragm movement to an electrical system, i.e. moving coil and ribbon microphones.

Dome – A specific shape of drive unit diaphragm intended to push and pull the air to create acoustic sound waves. Most tweeters use dome-shaped diaphragms which are driven around the circumference by the drive unit's motor system. 'Soft-domes' are made of a fabric – often silk – while metal domes are constructed from a light metal like aluminium, or some form of metal alloy.

Drive Unit/Driver – A physical device designed to generate an acoustic sound wave in response to an electrical input signal. Drive units can be designed to reproduce almost the full audio spectrum, but most are optimised to reproduce a restricted portion, such as a bass unit (woofer) or high-frequency unit (tweeter). A range of technologies are employed, with most being moving-coil units, but ribbon and electrostatic drive units also exist, each with a different balance of advantages and disadvantages.

Equaliser (cf. Filter) – A device which allows the user to equalise, balance or adjust the tonality of a sound source. Equalisers are available in the form of filters, shelf equalisers, parametric equalisers and graphic equalisers – or as a combination of these basic forms.

Equivalent Input Noise – A means of describing the intrinsic electronic noise at the output of an amplifier in terms of an equivalent input noise, taking into account the amplifier's gain.

FET – Field Effect Transistor. A solid-state semiconductor device in which the current flowing between source and drain terminals is controlled by the voltage on the gate terminal. The FET is a very high impedance device, which makes it highly suited for use in impedance converter stages in capacitor and electret microphones.

Fidelity – The accuracy or precision of a reproduced acoustic sound wave when compared to the electrical input signal.

Filter (cf. Equaliser) – Filters remove unwanted parts of the spectrum above or below a turnover frequency, and with the rate of attenuation versus frequency is called the slope. A high-pass (or low-cut) filter removes frequencies below the turnover frequency and usually has a slope of 6, 12 or 18dB/octave.

Filter Frequency – The 'turnover' or 'corner' frequency of a high- or low-pass filter. Technically, the frequency at which the signal amplitude has been attenuated by 3dB.

FireWire – A computer interface format based upon the IEEE 1394 standard and named FireWire by Apple computers (Sony's i.Link format is also the same interface). FireWire is a serial interface used for high speed isochronous data transfer, including audio and video. FireWire 400 (IEEE 1394–1995 and IEEE 1394a–2000) or S400 interface transfers data at up to 400Mb/s and can operate over cables up to 4.5metres in length. The standard 'alpha' connector is available in four and six-connector versions, the latter able to provide power (up to 25V and 8 watts). The FireWire 800 format (IEEE 1394b–2002) or S800 interface uses a 9-wire 'beta' connector and can convey data at up to 800Mb/s.

Flash Drive (cf. Solid-state Drive) – A large capacity solid-state memory configured to work like a conventional hard drive. Used in digital cameras and audio recorders in formats such as SD and CF2 cards, as well as in 'pen drives' or 'USB memory sticks'. Some computers are now available with solid state flash drives instead of normal internal hard drives.

Flutter Echoes – Short time-span sound echoes which can be created when sound waves bounce between opposite parallel walls in a small or moderately sized room. A shorter version of the 'slapback' echo which can be experienced in a larger hall when sound from a stage is reflected strongly from the rear wall.

Frequency Response – The variation in amplitude relative to the signal frequency. A measurement of the frequency range that can be handled by a specific piece of electrical equipment or loudspeaker, often expressed with a variation limit, such as –3dB.

General MIDI/GM – A universally agreed subset of the MIDI standard, created to enable manufacturers to build synthesizers, synth modules and plug-in instruments that exhibit an agreed minimum degree of compatibility.

Ground Loop/Ground-Loop Hum – A condition created when two or more devices are interconnected in such a way that a loop is created in the ground circuit. This can result in audible hums or buzzes in analogue equipment, or unreliable or glitch audio in digital equipment. Typically, a ground loop is created when two devices are connected together using one or more screened audio cables, and both units are also plugged into the mains supply using safety ground connections via the plug's earth pin. The loop is from one mains plug, to the first device, through the audio cable screen to the second device, back to the mains supply via the second mains plug, and round to the first device via the building's power wiring. If the two mains socket grounds happen to be at slightly different voltages (which is not unusual), a small current will flow around the ground loop. Although not dangerous, this can result in audible hums or buzzes in poorly designed equipment.

Ground loops can often be prevented by ensuring that the connected audio equipment is plugged into the same socket or mains distribution board, thus minimising the loop. In extreme cases it may be necessary to disconnect the screen connection at one end of the audio cables or use audio isolating transformers in the signal paths. The mains plug earth connection must NEVER be disconnected to try to resolve a ground loop problem as this will render the equipment potentially LETHAL.

GUI – Graphical User Interface (pronounced 'Gooey') – a software program designer's way of creating an intuitive visual operating environment controlled by a mouse-driven pointer or similar.

Hard Disk Drive (cf. Solid-state Drive) – The conventional means of computer data storage. One or more metal disks (hard disks) hermetically sealed in an enclosure with integral drive electronics and interfacing. The disks coated in a magnetic material and spun at high speed (typically 7200rpm for audio applications). A series of movable arms carrying miniature magnetic heads are arranged to move closely over the surface of the discs to record (write) and replay (read) data.

Headroom – The available 'safety margin' in audio equipment required to accommodate unexpected loud audio transient signals. It is defined as the region between the nominal operating level (0VU) and the clipping point. Typically, a high quality analogue audio mixer or processor will have a nominal operating level of +4dBu and a clipping point of +24dBu – providing 20dB of headroom. Analogue meters, by convention, don't show the headroom margin at all; but in contrast, digital systems normally do – hence the need to try to restrict signal levels to average around –20dBFS when tracking and mixing with digital systems to maintain a sensible headroom margin. Fully post-produced signals no longer require headroom as the peak signal level is known and controlled. For this reason it has become normal to create CDs with zero headroom.

High-range (high, highs) – The upper portion of the audible frequency spectrum, in a technical sense typically denoting frequencies above about 1kHz, but musically associated with frequencies over 3kHz.

Hub – Normally used in the context of the USB computer data interface. A hub is a device used to expand a single USB port into several, enabling the connection of multiple devices. Particularly useful where multiple software program authorisation dongles must be connected to the computer.

Hz/kHz – The standard abbreviation for Hertz (kilohertz) – a unit of measurement for frequency. 10Hz means ten complete cycles of a repeating waveform per second.

Impedance – The 'resistance' or opposition of circuit to the flow of current, when encountered in the context of electrical connections or the resistance that a medium presents to air flow, in the context of acoustics. Although measured in ohms, the impedance of a 'reactive' device such as a loudspeaker drive unit will usually vary with signal frequency and will be higher than the resistance when measured with a static DC voltage. Signal sources have an output impedance and destinations have an input impedance. In analogue audio systems the usually arrangement is to source from a very low impedance and feed a destination of a much higher (typically 10 times) impedance. This is called a 'voltage matching' interface. In digital and video systems it is more normal to find 'matched impedance' interfacing where the source, destination and cable all have the same impedance (e.g. 75 ohms in the case of S/PDIF).

Microphones have a very low impedance (150 ohms or so) while microphone preamps provide an input impedance of 1,500 ohms or more. Line inputs typically have an impedance of 10,000 ohms and DI boxes may provide an input impedance of as much as 1,000,000 ohms to suit the relatively high output impedance of typical guitar pickups.

Insert Points – The provision on a mixing console or 'channel strip' processor of a facility to break into the signal path through the unit to insert an external processor. Budget devices generally use a single connection (usually a TRS socket) with unbalanced send and return signals on separate contacts, requiring a splitter or Y-cable to provide separate send (input to the external device) and return (output from external device) connections. High-end units tend to provide separate balanced send and return connections.

Input Impedance – The input impedance of an electrical network is the 'load' into which a power source delivers energy. In modern audio systems the input impedance is normally about ten times higher than the expected source impedance – so a typical microphone preamp has an input impedance of between 1,500 and 2,500 ohms.

Isolator (also decoupler) – A device intended to prevent the transmission of physical vibration over a specific frequency range, such as a rubber or foam block.

Latency (cf. Delay) – The time delay experienced between a sound or control signal being generated and it being auditioned or taking effect, measured in seconds.

Lay Length – The distance along the length of a cable over which the twisted core wires complete one complete turn. Shorter lay lengths provide better rejection of electromagnetic interference, but make the cable less flexible and more expensive.

Limiter – An automatic gain-control device used to restrict the dynamic range of an audio signal. A Limiter is a form of compressor optimised to control brief, high level transients.

Loop – The process of defining a portion of audio within a DAW, and configuring the system to replay that portion repeatedly.

Loudspeaker (also Monitor and Speaker) – A device used to convert an electrical audio signal into an acoustic sound wave. An accurate loudspeaker intended for critical sound auditioning purposes.

Loudness – The perceived volume of an audio signal.

Low-range (low, lows) – The lower portion of the audible frequency spectrum, in a technical sense typically denoting frequencies below about 1kHz, but musically associated with frequencies below 300Hz.

Magnetic Shielding – Also called magnetic compensation (which is usually a more accurate description). A means of restricting the radiation range of the stray magnetic field from a drive unit's permanent magnet which might otherwise interfere with the correct operation of moving-coil meters or CRT television monitors. While it is possible to enclose a magnet in a soft-metal case to prevent a stray magnetic field this becomes very expensive for large magnets, and so a more common approach is to affix additional small external magnets with opposite polarities to cancel out the unwanted stray field.

Maximum SPL – The loudest sound pressure level that a device can generate or tolerate.

Metering – A display intended to indicate the level of a sound signal. It could indicate peak levels (e.g. PPMs or digital sample meters), average levels (VU or RMS meters), or perceived loudness (LUFS meters).

Mid-range (mid, mids) – The middle portion of the audible frequency spectrum, typically denoting frequencies between about 300Hz and 3kHz.

Microphone – A device used to convert an acoustic sound wave into an electrical signal.

MIDI – Musical Instrument Digital Interface. A defined interface format that enables electronic musical instruments and computers to communicate instructional data and synchronise timing. MIDI sends musical information between compatible devices, including the pitch, volume and duration of individual notes, along with many other aspects of the instruments that lend themselves to electronic control. MIDI can also carry timing information in the form of MIDI Clock or MIDI Time Code for system synchronisation purposes.

Mineral Wool – Made from natural or synthetic minerals in the form of threads or fibres tangled together to form a moderately dense 'blanket' which permits but impedes air flow and is useful in the creation of sound absorbers, often employed as a cheaper and more efficient alternative to polyurethane form.

Mirror Points – The positions on the walls or ceiling where, if the surface was covered with an optical mirror, one or both loudspeakers could be seen in the reflection. The mirror point is essentially any position on a boundary where sound waves from a sound source – usually a monitor loudspeaker – will be reflected directly to the listening position. This is therefore the ideal location to place an acoustic absorber to prevent audible reflections.

Mixer – A device used to combine multiple audio signals together, usually under the control of an operator using faders to balance levels.

Most mixers also incorporate facilities for equalisation, signal routing to multiple outputs, and monitoring facilities.

Modal Distribution – The characteristic distribution of low frequency resonances within a confined space such as a room.

Modes (room) – Specific patterns of low frequency sound reflection between surfaces in a room, resulting in resonant peaks.

Modelling – A process of analysing a system and using a different technology to replicate its critical, desired characteristics. For example, a popular but rare vintage signal processor such as an equaliser can be analysed and its properties modelled by digital algorithms to allow its emulation within the digital domain.

Monitor (also speaker and loudspeaker) – A device which provides information to an operator. Used equally commonly in the context of both a computer VDU (visual display unit) – such as an LCD screen – and a high quality, accurate loudspeaker intended for critical sound auditioning purposes.

Monitor Controller – A line-level audio signal control device used to select and condition input signals for auditioning on one or more sets of monitor loudspeakers. Some monitor controllers also incorporate facilities for studio talkback and artist cue mixes.

Mono – A single channel of audio.

Moving Coil – a technology used to convert energy between the mechanical and electrical domains. A coil of wire is allowed to move within a magnetic field. If the coil is caused to move it will generate an electric current proportional to the rate of movement (as in a microphone). If a varying electric current is passed through the coil it will move in proportion to the amount and direction of the current (as in a loudspeaker).

M-S (Mid-side) – A specialist form of coincident microphone array which, when decoded to left-right stereo, creates an equivalent XY configuration. In the MS array one microphone is pointed directly forward (Mid) while the second is arranged at 90 degrees to point sideways (Side). The Mid microphone can employ any desired polar pattern, the choice strongly influencing the decoded stereo acceptance angle. The Side microphone must have a figure-eight response and be aligned such that the lobe with the same polarity as the Mid microphone faces towards the left of the sound stage. Adjusting the relative sensitivity of the Mid and Side microphones affects the decoded stereo acceptance angle and the polar patterns of the equivalent XY microphones.

MTC – MIDI Time code – a format used for transmitting synchronisation instructions between electronic devices within the MIDI protocol.

Multi-timbrality – The ability of an electronic musical instrument to generate two or more sounds simultaneously.

Mutual Angle – the physical angle between two microphones, used to specify various microphone array configurations (e.g. 90 degrees for a Blumlein pair, or 110 degrees for an ORTF array).

Near-coincident – A means of arranging two or more directional microphone capsules such that they receive sound waves from the directions or interest at slightly different times due to their physical spacing. Information about the directions of sound sources is captured in the form of both level differences between the capsule outputs, generated by aiming directional polar patterns in different directions, and the timing differences caused by their physical spacing. Specific forms of near-coincident microphones include the ORTF and NOS arrangements.

Near Field – Describes a loudspeaker system designed to be used close to the listener. Some people prefer the term 'close field'. The advantage is that the listener hears more of the direct sound from the speakers and less of the reflected sound from the room.

NOS – A specific form of near-coincident microphone array devised by the Nederlandse Omroep Stichting (NOS), the Dutch national broadcaster. The technique employs a pair of small-diaphragm cardioid microphones mounted with a mutual angle of 90 degrees and spaced apart by 30cm. The theoretical stereo recording angle is 81°.

Ohm – Unit of electrical resistance.

Off-/On-axis – Directional microphones are inherently more sensitive to sound from one direction, and the direction of greatest sensitivity is referred to as the principle axis. Sound sources placed on this axis are said to be 'on-axis', while sound sources elsewhere are said to be 'off-axis'

Optimisation (of computer) – The concept of configuring a computer in such as way as to maximise its performance for certain tasks. In the context of a machine being used as a DAW, optimisation might involve disabling sub-programs that access the internet regularly or intermittently, such as email hosts, automatic program update checkers and so on. It might also include the structure of the hard drive, or the separation of program data to a system drive and audio data to a separate drive to minimise access times and maximise data throughputs.

ORTF – A specific form of near-coincident microphone array devised by the Office de Radiodiffusion Télévision Française (ORTF) at Radio France, the French national broadcaster. The technique employs a pair of small-diaphragm cardioid microphones mounted with a mutual angle of 110 degrees and spaced apart by 17cm. The theoretical stereo recording angle is 96 degrees.

Output Impedance – The effective internal impedance (resistance which many change with signal frequency) of an electronic device. In modern audio equipment the output impedance is normally very low. Microphones are normally specified with an output impedance of 150 or 200 ohms, although some vintage designs might be as low as 30 ohms.

Output Sensitivity – The nominal output voltage generated by a microphone for a known reference acoustic sound pressure level. Output sensitivity is normally specified for a sound pressure level of one Pascal (94dB SPL), and may range from about 0.5mV/Pa for a ribbon microphone, to 1.5mV/Pa for a moving coil, and up to 20 or 30mV/Pa for a capacitor microphone.

Overdubbing – Recording new material to separate tracks while auditioning and playing in synchronism with previously recorded material.

Passive Loudspeaker or Monitor – A loudspeaker which requires an external power amplifier, the signal from which is passed to a passive cross-over filter. This splits and filters the signal to feed the two (or more) drive units.

Patch (cf. Bank) – A specific configuration of sounds or other parameters stored in memory and accessed manually or via MIDI commands.

PCI Card – Peripheral Component Interconnect: an internal computer bus format used to integrating hardware devices such as sound cards. The PCI Local Bus has superseded earlier internal bus systems such as ISA and VESA, and although still very common on contemporary motherboards has, itself, now been superseded by faster interfaces such as PCI-X and PCI Express.

Phantom Power – A means of powering capacitor and electret microphones, as well as some dynamic microphones with built-in active impedance converters. Phantom power (P48) provides 48V (DC) to the microphone as a common-mode signal (both signal wires carry 48V while the cable screen carries the return current). The audio signal from the microphone is carried as a differential signal and the mic preamp ignores common-mode signals so doesn't see the common-mode power supply (hence the ghostly name, phantom). This system only works with a balanced three-pin mic cables. Two alternative phantom power specifications also exist, with P12 (12V) and P24 (24V) options, although they are relatively rare.

Phono plug (RCA-phono) – An audio connector developed by RCA and used extensively on hi-fi and semi-pro, unbalanced audio equipment. Also used for the electrical form of S/PDIF digital signals, and occasionally for video signals.

Pitch Bend – A means of temporarily changing the tuning of an audio signal generator, such as a synthesizer oscillator, either manually via a control wheel or under MIDI control.

Plug-in – A self-contained software signal processor, such as an Equaliser or Compressor, which can be 'inserted' into the notional signal path of a DAW. Plug-ins are available in a myriad of different forms and functions, and produced by the DAW manufacturers or third-party developers. Most plug-ins run natively on the computer's processor, but some require bespoke DSP hardware. The VST format is the most common cross-platform plug-in format, although there are several others.

Plug-in Power – Consumer recorders, such as MP3 recorders, are often equipped with a microphone powering system called 'Plug-In Power'. This operates with a much lower voltage (typically 1.5V) and is not compatible with phantom powered mics at all.

Polar Pattern – The directional characteristic of a microphone (omni, cardioid, figure-eight, etc.).

Polyphony – The ability of an instrument to play two or more notes of different pitches at the same time.

Pop Shield – A device placed between a sound source and a microphone to trap wind blasts – such as those created by a vocalist's plosives (Bs, Ps and so on) – which would otherwise cause loud popping noises as the microphone diaphragm is over-driven. Most are constructed from multiple layers of a fine wire or nylon mesh, although more modern designs tend to use open-cell foam.

Power Amplifier – A device which accepts a standard line-level input signal and amplifies it to a condition in which it can drive a loudspeaker drive unit. The strength of amplification is denoted in terms of Watts of power.

Powered Loudspeaker or Monitor – A powered speaker is a conventional passive loudspeaker but with a single power amplifier built in or integrated with the cabinet in some way. The amplifier drives a passive crossover, the outputs of which connect to the appropriate drive units.

Pre-amp – Short for pre-amplification: an active gain stage used to raise the signal level of a source to a nominal line level. For example, a microphone pre-amp.

Project Studio – A relatively small recording studio facility, often with a combined recording space and control room.

Proximity Effect – Also known as 'Bass tip-up'. The proximity effect dramatically increases a microphone's sensitivity to low frequencies when placed very close to a sound source. It only affects directional microphones – omnidirectional microphones are immune.

Quantisation – Part of the process of digitising an analogue signal. Quantisation is the process of describing or measuring the amplitude of the analogue signal captured in each sample, and is defined by the wordlength used to describe the audio signal – e.g. 16 bits.

Rack Mount – A standard equipment sizing format allowing products to be mounted between vertical rails in standardised equipment bays.

RAM – Random Access Memory: the default data storage area in a computer, normally measured in Gigabytes (GB).

Reflection – The way in which sound waves bounce off surfaces.

Reverb – Short for reverberation. The dense collection of echoes which bounce off of acoustically reflective surfaces in response to direct sound arriving from a signal source. Reverberation can also be created artificially using various analogue or, more commonly, digital techniques. Reverberation occurs a short while after the source signal because of the finite time taken for the sound to reach a reflective surface and return – the overall delay being representative of the size of the acoustic environment. The reverberation signal can be broadly defined as having two main components, a group of distinct 'early reflections' followed by a noise-like tail of dense reflections.

Reverb Decay – The time taken for sound waves reflecting within a space to lose energy and become inaudible. A standard measurement is 'RT_{60}' which is the time taken for the sound reflections to decay by 60dB.

Ribbon – A type of electromagnetic microphone in which the diaphragm is also an electrical conductor which is placed within a strong magnetic field. As the diaphragm moves it generates a small proportional current.

Satellite (speaker) – Normally used in the context of a loudspeaker system employing a subwoofer to reproduce the lowest frequencies, with smaller 'satellite' loudspeakers to reproduce the higher frequencies.

Self Noise – A term used to describe the electronic noise contribution of the active impedance converter in an electrostatic microphone, and specified in terms of the equivalent acoustic sound pressure level required to create the same signal voltage as the amplifier's noise floor. The self-noise figure dictates the lower limit of the microphone's total dynamic range.

Sensitivity – The efficiency of a loudspeaker in converting an electrical input signal to an acoustic output signal, or of a microphone converting sound to an electrical signal.

Sample – Either a defined short piece of audio which can be replayed under MIDI control; or a single discrete time element forming party of a digital audio signal.

Sample Rate (cf. Bit Rate) – The rate at which a digital audio signal is intended to operate, normally denoted either in terms of kilo-samples per second (kS/s) or kilo-Hertz (kHz). The audio bandwidth must be less than half the sample rate, which is high quality audio systems operate at 44.1 or 48kS/s to provide an audio bandwidth of at least 20kHz.

Sampler – A hardware device or software program which replays (and possibly captures) short audio excerpts under MIDI control.

Sequencer – A device which records and replays MIDI instructions. Original sequencers were hardware devices but most are now software and are integrated into most DAWs.

Shockmount – a mechanical isolator intended to prevent the transfer of vibrations which may be transmitted through a microphone stand from reaching a microphone where they would otherwise produce unwanted low frequency sound.

SMPTE TimeCode – A means of affording recordings with reliable positional information coded to resemble clock time, originally used to identify individual picture frames in video and film systems.

Solid-state Drive (cf. Hard Disk Drive) – A large capacity solid-state memory configured to work like a conventional hard disk drive. Some computers are now available with solid-state flash drives instead of normal internal hard disk drives. Also used in digital cameras and audio recorders in formats such as SD and CF2 cards, as well as in 'pen drives' or 'USB memory sticks'.

Sound Card – A dedicated interface to transfer audio signals in and out of a computer. A Sound Card can be installed internally, or connected externally via USB2 , USB3, Firewire or Thunderbolt. They are available in a wide range of formats, accommodating multiple analogue or digital audio signals (or both) in and out, as well as MIDI data in and out.

Soundproofing – The use of materials and construction techniques with the aim of preventing unwanted sound from entering or leaving a room.

Spaced Array – A means of arranging two or more microphone capsules such that they receive sound waves from different directions at different times – these timing differences being used to convey information about the relative directions of those sound sources. The technique is usually used with omnidirectional microphones, although directional mics can also be employed. The best known form of spaced array is the Decca Tree. Mono-compatibility is often reduced because the timing differences between channels often results in comb-filtering colouration when the channels are summed to mono.

S/PDIF – Sony/Philips Digital Interface. Pronounced either 'S-peedif' or 'Spudif'. A stereo or dual-channel self-clocking digital interfacing standard employed by Sony and Philips in consumer digital hi-fi products. The S/PDIF signal is essentially identical in data format to the professional AES3 interface, and is available as either an unbalanced electrical interface (using phono connectors and 75ohm coaxial cable), or as an optical interface called TOSlink.

Speaker (also Loudspeaker and Monitor) – An accurate loudspeaker intended for critical sound auditioning purposes.

SPL – Sound Pressure Level. A measure of the intensity of an acoustic sound wave. Normally specified in terms of Pascals for an absolute value, or relative to the typical sensitivity of human hearing. One Pascal is 94dB SPL, or to relate it to atmospheric pressures, 0.00001 Bar or 0.000145psi!

SRA – See **Stereo Recording Angle**

Standing Waves – Resonant low frequency sound waves bouncing between opposite surfaces such that each reflected wave aligns perfectly with previous waves to create static areas of maximum and minimum sound pressure within the room. (See also **Modes and Modal Frequencies**)

Stereo – By convention, two channels of related audio which can create the impression of separate sound source positions when auditioned on a pair of loudspeakers or headphones.

Stereo Recording Angle – The angle over which sound sources can be captured by a microphone array. For a stereo array with a stereo recording angle of 90 degrees, sound sources can be placed ±45 degree relative to the array's centre front axis, with a source at the extreme angle appearing fully left or right in the stereo image.

Subwoofer – A specific type of efficient loudspeaker system intended to reproduce only the lowest frequencies (typically below 120Hz).

Surround – The use of multiple loudspeakers placed around the listening position with the aim of reproducing a sense of envelopment within a soundstage. Numerous surround formats exist, but the most common currently is the 5.1 configuration in which three loudspeakers are placed in front of the listener (at ±30degrees and straight ahead), with two behind (at ±120 degrees or thereabouts), supplemented with a separate subwoofer.

Synthesis – The creation of artificial sound.

Synthesizer – A device used to create sounds electronically. The original synthesizers were hardware devices and used analogue signal generation and processing techniques, but digital techniques took over and most synthesizers are now software tools.

Talkback – A system designed to enable voice communication between rooms, such between an engineer and producer in a control room and performers in an adjacent recording room.

TRS – A type of quarter-inch jack plug with three contacts (Tip, Ring and Sleeve), used either for stereo unbalanced connections (such as on headphones) or mono balanced connections (such as for line-level signals). Physically compatible in size with the TS quarter-inch jack plug used for electric guitars and other instruments.

Tube – See **Valve**

Tweeter – The colloquial term to describe a loudspeaker drive unit optimised for the reproduction of high frequencies. (See **Woofer**).

USB – Universal Serial Bus. A computer interface standard introduced in 1996 to replace the previous standard serial and parallel ports more commonly used. The original USB1.0 interface operated at up to 12Mb/s, but this was superseded in 2000 by USB2.0 which operates at up to 480Mb/s. Most USB interfaces can also provide a 5V power supply to connected devices. USB3.0 was launched in 2008 and is claimed to operate at rates up to 5Gb/s, but it is only now (2011) starting to appear on hardware.

Valve – Also known as a 'tube' in America. A thermionic device in which the current flowing between its anode and cathode terminals is controlled by the voltage applied to one or more control grid(s). Valves can be used as the active elements in amplifiers, and because the input impedance to the grid is extremely high they are ideal for use as an impedance converter in capacitor microphones. The modern solid-state equivalent is the Field Effect Transistor or FET.

Vibrato – A cyclical variation in pitch often employed in musical performance.

W (Watt) – Unit of electrical power.

-way (as in, 2-way, 3-way) – A colloquial way of denoting how many separate frequency bands are reproduced by a loudspeaker. Most are two-way systems with a woofer and tweeter, but some are three way with a woofer, midrange and tweeter.

XLR – A very robust and latching connector commonly used to carry balanced audio signals such as the outputs from microphones or line level devices. An XLR is a type of connector developed by US manufacturer, Cannon, and used widely in professional audio systems. The company's original X-series connector was improved with the addition of a latch (Cannon XL) and a more flexible rubber compound surrounding the contacts to improve reliability (Cannon XLR). The connector format is now is available in numerous configurations, from many different manufacturers, and with several different pin configurations. Standard balanced audio interfaces – analogue and digital – use three-pin XLRs with the screen on pin 1, the 'hot' signal on pin 2 and the 'cold' signal on pin 3.

XY – A specific way of mounting two directional microphone capsules such that they both receive sound waves from any direction at exactly the same time. Information about the direction of a sound source is captured in the form of level differences between the two capsule outputs. Commonly, the two microphones in an XY array are mounted with a mutual angle of 90 degrees, although other angles are sometimes used. The two capsules will have the same polar pattern, the choice of which determines the stereo recording angle (SRA). The XY configuration is entirely mono-compatible because there are no timing differences between the two channels.

Index

acoustic foam 14–15, 21–4, 33, 38, 261
acoustic guitars 75
  figure-of-eight microphones 138–9
  improving DI'd guitar 144–6
  microphones 136–8
  mixing 231
  optimum mic position 139–42
  recording environment 135–6
  stereo recording 142–4
  undersaddle piezo-based pickup/preamp 133–4
acoustic piano 75
acoustic screens 01–2, 70
acoustic treatment 20, 24, 27, 33, 38, 261
active loudspeaker (or monitor) 45, 261
air EQ 124, 223, 224
air-conditioning 92
amp/amplifier 261
  miking a guitar amp 150–1
  modelling preamps 148–50
  re-amping 163
  remote 162–3
  types 154
analogue 45, 111–12, 239, 261
arming 261
audio interface 45, 106, 108, 126, 137, 156, 163, 205, 261–2
automatic tuning 225–7
automation 185, 199, 210, 223, 225, 240–1, 262
auxiliary sends 204–6, 262

back electret 116–17, 158, 263
balanced/unbalanced cables 99–103, 262–3

barrier mat (or sheet rock, dead-sheet) 34–5, 90
bass guitar 166–9, 170, 227–8
bass problems 9–10, 48
bass response 10, 12, 39, 48, 49–50, 167, 247, 263
bass tip-up see proximity effect
bass traps 32, 48, 263
  broadband 32–4
  foam 33
  loft insulation 33–4
  mattresses/sofas 33
  mineral wool 33
  soft-furnishings 37
  stud partition/dry walls 37
  windows/doors 37
bit rate 263
  see also sample rate
Blumlein array 263–4
boom 56, 176, 264
boundary 43, 122, 264
bracketing 221–3
buffer 73, 126, 156, 264
bussing 201–4

cabinet 1, 3, 45, 46, 47, 73, 148, 151, 152, 227, 264
cabinet resonance 264
cable screens 103–5
cables/connections 183
  balanced/unbalanced 99–103, 262–3
  cable screens 103–5
  case study 97–9
  cleaning 114
  cold/hot signal wire 101–2
  disconnecting 96–7
  examples 93–7
  gain structure 107–10

cables/connections (Cont.)
  impedance 101, 110–11
  on the level 111–12
  speaker cables 113–14
  summary 114–15
  unbalanced to balanced
    interfacing 105–6
capacitor 75, 110, 116–18, 122,
  136–7, 158–9, 177, 182,
  183–4, 264
capsule 117, 123, 124, 264
CD-burning 256–9
CD-R 239, 256
ceilings 90, 120
channel fader 128
Clapton, Eric 150
click track 265
clipping 40, 107–8, 132, 157,
  200, 236–7, 249, 253, 265
clocking 265
clothes closet 63–5, 73
coincident 142, 143, 179, 265
colouration 5, 62, 63, 71,
  75, 265
coloured sound 3–5
comfort reverb 126–8
communication with
  musicians 72
comping 266
compression 252–3
  parallel 254–5
compressor 131, 186,
  209, 266
cone 42, 46, 47, 160, 161, 167,
  223, 266
corner traps 32–4, 121
CPU 266
crossover 266
crossover frequency 266
CRT (cathode ray tube) 166

D-A converter 200, 250
daisy chain 266–7
DAT machine 239
DAW (Digital Audio Workstation)
  72, 73, 108, 127, 144, 179,
  196–7, 201–4, 237, 267
dB (deciBel) 267
dead zone 59
Decca tree 267

decoupler (or isolator) 14, 83,
  87, 91, 267
delay 268
DI box 74–5, 145, 146, 166,
  169, 227–8, 268
diaphragm 75, 116, 122, 123,
  136–8, 160, 167, 183–4,
  268
digital 268
digital-modelling 148–50
distortion 236–7
dome 268
doors 84, 86–7
double-lapped screen 268
drive unit/driver 269
drums 71, 171–3
  close miking 182–4
  damping 173
  divide and conquer 189–90
  isolated from floor 91
  kick drum 176–8
  loops 191–2
  maintaining 171
  mic positioning 178–82
  overheads 179
  pre-mixing and panning
    186–7
  replacement 190–1
  room environment 171–2
  sampled 187
  snares 175–6
  summary 192
  tracking 185–6
  tuning the kit 173–4
dummy loads 153
duvets 62, 70, 73, 119–21
dynamic 268
dynamic processors 250–2

editing 240–1, 255–6
electric guitar 148
  amplifier types 154
  bass guitar 166–9
  case study 152, 157–8
  dummy loads/speaker
    emulation 153
  hiss and hum 163–4, 166
  microphones 157
  miking a guitar amp 150–1
  mixing 229–31

electric guitar *(Cont.)*
  model behaviour 148–50
  noise-reducing modifications
    164–5
  power soak units 151–2
  re-amping 163
  reality check 169–70
  remote amping 162–3
  software emulations 154–7
  sound styles 158–62
electromagnetic interference 101
EQ 124, 125, 126, 144–5,
  166, 183, 211, 220–3,
  224, 237, 246, 248–50
equaliser 141, 144, 220,
  224, 269
equivalent input noise 269

faders 128, 210
Fender, Leo 148, 150
FET (Field Effect Transistor) 269
fidelity 269
filter 63, 96, 121, 124, 139,
  187, 221, 223, 224,
  228, 230, 231, 236–7,
  248, 269
filter frequency 269
Fingerprint EQ 144
FireWire 209–70
flanking transmission 91
flash drive 270
  see also solid-state drive
flutter echo 30, 270
frequency response 3–5, 270

gain structure 40–2, 107–10
general MIDI/GM 270
ground loop/ground loop hum
  95–6, 270–1
groups 206–7
GUI (graphical user interface)
  271

hard disk drive 271
headroom 74, 107–10, 112,
  132, 149, 186, 271
high-range (high, highs) 271
hiss and hum 163–4, 166
hub 271
Hz/kHz 271

impedance 110–11, 271–2
input impedance 111, 272
insert points 217, 254, 272
Internet audio 259–60
isolator 272

keyboards, real/virtual 73–5
kick drum 176–8

latency 154–6, 272
lay length 272
LCD 166
Led Zeppelin 150
level adjustments 231–4
lighting 166
limiter 232–3, 250–1, 273
limp-mass absorbers 34–6
loop 95–6, 270–1, 273
lossy materials 81, 83, 90
loudness 273
loudspeaker (monitor or
  speaker)
  active/passive 44–6
  bass problems 9–10
  frequency response 3–5
  phasing 42–3
  placement 1–2, 11–12
  positioning/angle 10–13, 39,
    46 8
  stands 13–15, 91
  suitable 44–6
low-range (low, lows) 273

M-S (mid-side) 275
magnetic shielding 273
Match EQ 144
Marshall stacks 148
mastering
  compression 252
  development of 242–4
  dynamics 250–2
  editing tasks 255
  EQ 248–50
  evaluation 246–7
  final checks 256–7
  gently does it 252–3
  home vs pro 245–6
  Internet audio 259–60
  parallel compression 254–5
  process 247–8

maximum SPL 273
metering 108, 244, 273
microphones (mics) 273
  analogue 111
  capacitor 117, 136–7
  cardioid pattern 119, 137
  choosing 125–6, 157
  close/distant 71, 72, 124,
    161–2, 182–4
  combinations 158–9
  dynamic 75
  figure-of-eight 138–9
  hi-hat 183–4
  kick mic position 177–8
  miking a guitar amp 150–1
  omni-pattern 137–8
  optimum position 139–42,
    178–80, 182
  overhead 71
  power soak units 151–2
  powering 117–19
  presence peak 126
  ribbon 137, 159–60, 177, 279
  shaping sound via position
    160–2
  small-diaphragm capacitor 75
  talkback 72, 282
  three-mic technique 182
  tube 117
mid-range (mid, mids) 273
Mid-Side 143
MIDI 73, 147, 187, 192, 273–4
mineral wool 24–6, 33, 58, 274
mirror points 53–6, 274
mix bus 74
mixer 274
mixing 193–5
  acoustic guitars 231
  art of 208–9
  automatic tuning 225–7
  auxiliary sends 204–6
  bass guitar 227–8
  buses and sends 201–4
  case study 214–15, 220–1,
    229, 237–8
  combining mix automation
    with advanced editing
    240–1
  distortion 236–7
  electric guitars 229–31

mixing (Cont.)
  EQ bracketing 221–3
  EQ presets 214–15
  EQ to separate 220–1
  getting started 196–7
  grouping the VCA way 207–8
  initial balance 201
  judging the balance 231–4
  keep it clean 195
  medium 239
  noise gate 234–6
  panning 215–17
  perspective 218–20
  plug-in presets 212–14
  polishing and shaving 209
  preliminary housekeeping
    197–8
  problem solving 234
  reverb space 217
  sends and groups 206–7
  simply the wrong sound 237
  track 'cleaning' 199–201
  vocal levelling 209–12
  vocal reverbs 225
  vocals 223–5
mixing console 45, 111
modal distribution 274
modelling 134, 144, 148–50, 274
modes (room) 274
monitor 274
  see also loudspeaker (monitor
    or speaker)
monitor controller 274
monitoring 1
  artist 129–30
  checking bass problems
    9–10, 48
  do's and don'ts 60
  frequency response 3–5
  magic triangle 13
  perfect symmetry rule 12–13
  placement of speakers 1–2,
    11–12
  pointing down long room
    dimension 10
  practical solutions 39–59
  room modes 5–7
  room reflections 16–17
  shape/size of room 7–9
  stands 13–15

mono 275
moving coil 137, 159, 275
MTC (MIDI time code) 275
multi-timbrality 275
mutual angle 275

near field 275–6
near-coincident 275
noise gate 234–6
NOS 276

off/on-axis 3–4, 256
ohm 276
optimisation (of computer) 276
Orange Book 256
ORTF 143, 276
output impedance 111, 276
output sensitivity 276
overdubbing 129, 277
OVU 108

panning 215–17, 218
passive loudspeaker or monitor
    42, 44–5, 46, 113, 277
patch 277
PCI (peripheral component
    interconnect) card 277
PFL (Pre Fade Listen) 108
phantom power 16, 117–18, 277
phono plug (RCA-phono) 277
pitch bend 277
plug-in 148–50, 170, 179, 237,
    277–8
  power 278
  preset 212–14
polar pattern 3, 278
polyphony 278
pop shield 123–4, 223, 278
post-fade 205–6
power amplifier 278
power soaks 152
powered loudspeaker (or
    monitor) 278
pre-amp 133–4, 278
pre-fade 204–5
presets 157
project studio 13, 59, 61, 69,
    77, 84, 93–4, 102, 114,
    133, 137, 151, 178, 193,
    237, 278

proximity effect (or bass tip-up)
    124, 137, 138, 139, 159,
    278
PZM (Pressure Zone
    Microphones) 180, 182

quantisation 279

rack mount 279
RAM 279
recording space
  acoustics 61–3
  different rooms/spaces 72–3
  instruments 69–70
  keyboards (real/virtual) 73–5
  reflections/resonances 71
  vocal booths 63–9
Red Book 257–8
reflection 279
reflective areas 180
  angled wall panels 31
  CDs as reflectors 28, 31, 55
  floors 71
  mirror analogy 31
  non-parallel walls/shaped
    ceilings 31–2
resonances 71
resonant modes 5–7
reverb 186–7, 217, 225, 279
reverb decay 279
ribbon 137, 159–60, 177, 279
right-brain activity 197
room
  modes 274
  rear walls 57–60
  reflections 16–17
  shape/size 7–9
  sound resonances 5–7, 20
  symmetry 11–12
  width/length effect 10

S/PDIF 103, 281
sample 280
sample rate 256, 280
  see also bit rate
sampler 18, 61, 220, 280
satellite (speaker) 279
scattering 30–2
self noise 279
sends 204–7

sensitivity 41, 157, 280
sequencer 280
shockmount 280
sine wave test sequence
    18–19, 48
SMPTE TimeCode 280
snare drum 175–6, 182
software emulations 154–7
solid-state drive 280
    see also flash drive
sound
    isolation 77–8, 92
    speed of transmission
        79–80
    vibrational energy 78
    wavefront 79
sound absorbers 20
    acoustic foam 22–4
    bass traps 32–3
    carpet 27–8
    commercial products 38
    corner traps 32–4
    curved 120
    decay time 20, 29
    diffusing/reflecting sound
        30–2
    duvets 62, 70, 119–21
    egg box myth 30
    expanded polystyrene
        panels 29
    free bass absorption 37
    limp-mass 34–6
    mineral wool 24–6, 58
    mirror points 53–6
    open-plan-office dividing
        screens 29–30
    porous 20–1
    positioning 52–6
    role 52
    studio sofa (bed, mattress)
        33, 56, 179
sound card 280
sound leaking 77
    ceilings 90
    complaints 83
    doors 84, 86–7
    floors 88–90
    masked by daytime noise 85,
        86–7
    vibration-borne sound 90–1

sound leaking *(Cont.)*
    walls 79–84
    windows 84, 85–6
soundproofing 77–8, 280
spaced array 281
speaker 281
    see also loudspeaker (monitor
        or speaker)
speaker cables 113–14
speaker emulation 153
'The Spherical Bermuda Triangle
    of Death' 8
SPL (sound pressure level) 281
SRA (stereo recording angle) 281
standing waves 5
Stratocaster guitar 148
studio vibe 128
subwoofer 281
    2.1 system 49–50
    calibration 51–2
    level, phase, cut-off
        frequency 51
    positioning 50
surround 282
sweet spot 71
synthesis 282
synthesiser 222, 282

talkback 72, 282
tape op 196
target sound 144
templates 128
track cleaning 199–201
TRS (tip-ring-sleeve) 105, 282
TS (tip-sleeve) 106
tub combo 73
tube *see* valve
tuned traps 32
tweeter 13, 14, 16, 46, 47,
    247, 282

USB (Universal Serial Bus) 282

valve 117, 282
vibrato 283
vocal booths 121
    clothes closet example 64–5
    size 63–4
    sound absorption 63–4
    treatment 65–7, 69

vocal recording 116–17
  artist monitoring
     129–31
  backing vocals 224–5
  boxy-sounding 118
  case study 122
  comfort reverb 126–8
  directionality 119–21
  mic choices 125–6
  mic powering 117–19
  position 122–4
  processing prior to
     recording 131–2
  studio 'vibe' 128
  summary 132
VU meter 108

W (watt) 283
walls
  adding mass 80–1
  air gap 80, 82–3
  cavity wall 80
  dry wall/studding 84
  lossy layers 81, 83
  plasterboard-on-studding 82, 83
  silicone rubber/mastic 82–3
  vibrational energy 80
WAV 256
way (as in 2-way or 3-way) 283
windows 84, 85–6

X-Y 143, 179, 283
XLR 64, 102, 105, 106, 283

Milton Keynes UK
Ingram Content Group UK Ltd.
UKHW020844141024
449569UK00003B/67

9 781138 468894